To Iain Milne,
i thanks a best
wishes Michael Lee

MACADAM

Company Histories:
Men and Machines: D. Napier & Son 1808–1958
(with C. H. Wilson)
Hard Roads and Highways: SPD Limited 1918–1968
The Weir Group: a Centenary History
Imperial Chemical Industries: a History
Vol. 1 *The Forerunners, 1870–1926*
Vol. 2 *The First Quarter-Century, 1926–1952*
Metal Box: A History
A House in the City
Fifty Years of Unilever

Biography:
Architect of Air Power: the Life of the first Viscount Weir, 1877–1959

General History:
Life in Victorian England (reissued as *Victorian England*)
*Professional Men: the Rise of the Professional Classes
in Nineteenth-Century England*
The Middle Classes
Call to Arms: the Great Volunteer Army of World War I
Macadam: the McAdam Family and the Turnpike Roads

Frontispiece – John Loudon McAdam (1756–1836) by an unknown
   (over)   artist. In the middle distance, left of centre, there is
            a road gang.

National Portrait Gallery, London

MACADAM

*The McAdam Family and
the Turnpike Roads
1798–1861*

W. J. Reader

HEINEMANN:LONDON

William Heinemann Ltd
10 Upper Grosvenor Street, London W1X 9PA
LONDON   MELBOURNE   TORONTO
JOHANNESBURG   AUCKLAND

First published in Great Britain 1980
© W. J. Reader and Tarmac Limited 1980

SBN 434 62502 7

Printed and bound in Great Britain by
Morrison & Gibb Ltd, London and Edinburgh

To Ann

Acknowledgements

The line drawings in the text, reproduced by courtesy of the Curator, Luton Museum, are by W H Pyne (1769–1843). They were published, with text by C Gray, in *Microcosm, or a Picturesque Delineation of the Arts, Agriculture, and Manufactures of Great Britain*, which appeared in parts starting in 1803, and in a complete edition. I am grateful also to the owners of the other illustrations for permission to reproduce them. Individual acknowledgement has been made in the captions.

By courtesy of The Goodwood Estate Company and with acknowledgements to the West Sussex Record Office, I am enabled to quote extracts from papers belonging to The Goodwood Estate. I am grateful to the Marquis of Salisbury for access to the archives of Hatfield House and to Mr Harcourt Williams for help in consulting them.

The extract from *The Wind in the Willows* by Kenneth Grahame on page 205 is reproduced by courtesy of Methuen Childen's Books, Text copyright Unversity Chest, Oxford.

Contents

Contents

Illustrations

Illustrations

One of the gates belonging to the Bath Turnpike Trust, 1826.
A stage coach at a turnpike gate, evidently on a wet evening.

Foreword

I HOPE THIS book will fill a gap in studies of the Industrial Revolution. A biography of J L McAdam, by his great-great-granddaughter Mrs Pember ('Roy Devereux'), was published in 1936 and another was prepared as a PhD thesis by Dr Robert H Spiro Jr in 1950, but never published. So far as I am aware, no one until now has looked at the work of J L McAdam, his sons and grandsons, as a whole, yet it was undoubtedly of importance in the development of what Professor Peter Mathias has called 'The First Industrial Nation'.

I am extremely grateful to Tarmac Limited, whose generosity has made the book possible although the firm has no connection with any branch of the McAdam family, did not come into existence until forty years after their work on roads had ceased, and makes no appearance in the book until very nearly the last page. J F Earle, of Tarmac, has given me congenial and stimulating support throughout, including invaluable aid in navigating the turnpike roads of Somerset with a map 150 years old.

In the early stages of the work I approached Mrs M C Bowker, the Revd A H Purcell Fox, Mark Lawrence and Cecilia Stride for family information. I should like to thank them for their interest and, in certain cases, for the loan of documents. Sue Gillies, Reference Librarian of the New York Historical Society, guided me to material in the Society's collection.

For help with Parliamentary Papers and certain other material I have relied on the staff of the Guildhall Library, London, and

I have made widespread use of turnpike trust papers in county record offices. I am grateful for the courteous interest shown by twenty-eight county archivists, and in particular to the archivists and their staff in ten counties (including the staff of Sheffield Central Library) where documents relating to the work of various members of the McAdam family are preserved, as shown in Appendix Two. Permission to reproduce material for illustrations is separately acknowledged.

The last chapter, covering the years between the stage coach and the motor car, relies heavily on work by Lauraine Dennett, to whom I am deeply indebted.

Sylvia Southcombe and Kathy Meek have been kind enough to read and comment upon the text in typescript.

The British Almanac, Samuel Tymms' epitome and continuation of *Camden's Britannia*, and Harry Inglis's *'Contour' Road Book of England* contain a great deal of information and period atmosphere not otherwise readily accessible. For them, and for a great deal else besides of a similar nature, I am indebted to my wife's assiduity in the discovery and exploration of invaluable centres for historical research : second-hand bookshops.

London, April 1979 W J Reader

Postscript

As all sums of money quoted in this book date from pre-decimalization times, it may help to remember that there were twelve old pennies in one old shilling (i.e. 12d = 1s) and that one old shilling equals five new pence (i.e. 1s=5p). Consequently, ten old shillings equal fifty new pence (i.e. 10s = 50p), and twenty old shillings (20s) or one hundred new pence (100p) equal one pound (£1).

I

The Roads of the Regency

J OHN LOUDON MCADAM (1756–1836), like another Scot, his
younger contemporary Charles Macintosh (1766–1843), gave
his name to the language through his work. From about the
year 1820 onward the word 'macadam' came to mean, for
most people, a hard smooth road surface over which carriages
and wagons could travel as fast as horses could be made to
draw them. During the reigns of King George IV (1820–1830)
and King William IV (1830–1837) England rapidly acquired
much the best road system seen in the country since the fall of
the Roman Empire: a system over which mail coaches could
travel at average speeds of up to ten miles an hour or more[1],*
and lighter vehicles faster. Then, in the first few years of Queen
Victoria's reign, the improvement and even the maintenance of
long-distance roads ceased as quickly as the railways spread,
for road transport, apart from purely local journeys, seemed to

* See relevant reference notes at the end of each chapter.

1

have become obsolete. It was the triumph of John Loudon McAdam to play a large part in bringing the roads of the 1820s and the 1830s into being, and it was the tragedy of his sons, particularly Sir James McAdam (1786–1852), to see their life-work nullified by the roads' decline.

'Macadam', as conceived of by J L McAdam, consisted simply of a layer of stone, broken small, resting directly on the sub-soil. 'Every piece of stone put into a road', said McAdam in 1811, 'which exceeds an inch in any of its dimensions is mischievous'; later he said he obliged surveyors working for him to carry pocket scales so that they could weigh sample stones, any weighing more than about six ounces being almost certainly too large. Stones broken to the right size were angular, not round, and under the weight of traffic they consolidated, without tar or any other binding material, to provide the surface required.[2]

If, against McAdam's advice, often and strongly repeated, earth or sand was applied as a binder, the results might be chokingly spectacular. On 18th April 1905 the *Daily Mail* reported that on the Holyhead road near Wolverhampton a car had run into a carrier's cart. The motorist pleaded not only that he was unable to see any vehicle approaching but that the pall of dust was so dense that he was unaware of being on the wrong side of the road. The *Mail*'s leader writer concluded, no doubt rightly, that 'the main principles which the famous Macadam taught' had been abandoned, and earth had been dumped on the road to bind the surface.

The comment shows how long McAdam's 'main principles' were remembered: from the stage coach to the motor car. There was far more to them, however, than his method of surfacing a road. His achievement, as he and his sons were well aware, comprehended far more than that. He and they developed a complete system of road management, covering both the technique of construction and directions for selection and training of staff and labour, contracting for materials and services, financial control and accounting.

Neither McAdam nor any of his sons was an engineer and

McAdam himself expressly disclaimed the title before a parliamentary committee in 1819.[3] McAdam described himself, his sons, and those whom he and they trained as 'road surveyors'. Their function was administrative and managerial: to take charge of money, labour and materials and to see that they were applied to road improvement and maintenance with the least waste and to the greatest effect. The surveyor's job was to create and preserve a good road surface. As well as that it was a matter of straightening here, widening there, reducing gradients, removing obstacles and, apart from a few miles every now and then, it was not a matter of laying out new routes. J L McAdam and his sons – William (1781–1836), (Sir) James Nicoll (1786–1852) and Loudon (1798–1857) – were not constructors on a grand scale like Thomas Telford (1757–1834) or the early railway engineers and contractors, but they deserve to be ranked among the very earliest of modern professional managers.

J L McAdam, according to his own account,[4] began to take an interest in roads in the 1780s, but it was not until after he moved from Scotland to Bristol, in 1798, that his hobby began to turn into a business, and that, as we shall see later, only happened gradually. The roads he began to examine were the roads of a country in which modern industrial society was beginning to take shape. Steam power was nothing new and scores of hopeful inventors, mostly self-taught – for who else was to teach them? – were trying to extend its applications, not only in pumping and other stationary machinery but in boats, locomotives and road carriages.[5] Intrepid tourists who penetrated North-Western England could see, in cotton mills, an early stage in the organization of large-scale production. It was appearing also in a few other trades, such as iron-working, brewing and shipbuilding, but in general the British economy was still based on small employers, hand-craftsmanship and outwork, rather than on large factories and mechanized production, and so it would remain for a long time to come. In the business community the merchant, not the manufacturer (the word still meant principally a craftsman), was king – especially the cotton merchant. Cotton goods, produced in rapidly growing abundance, were already by 1800 much the largest item in Great Britain's export trade, as they would remain until well into the 20th century. The export trade itself, the greatest in the world and supported by an array of service trades – shipping, insurance, banking, finance – showed how far ahead of any other country, in matters of commerce and industry, Great Britain already stood: not that this would have been very obvious to an observer, except in the manufacturing districts, the coalfields, the great ports and the City of London.

Most of Great Britain's rapidly growing population – 10.7 millions in 1801, 12.15 millions in 1811, 14.2 millions in 1821 – lived on and by the land, and it was not until the census of 1851 that more people were reported in towns than in the country-side. The wealthiest men in the country, by far, were not factory owners, merchants, bankers, or businessmen of any kind, but great landed proprietors like the Dukes of Bedford, Bridgewater, Devonshire, Northumberland, each of whom as early as 1790 'could comfortably dispose of over £50,000 a year'.[6] At the other end of the scale, farm labourers were lucky to earn a few shillings a week while farming was prosperous during the war but found it difficult to get work at all – unless on the roads – in the desperately hard times that followed Waterloo.

Gross inequalities of this kind fed revolutionary aspirations, which were crushed with ruthless vigour. The war, while it lasted, brought inflation and scarcities: after it stopped, depression, disorder, and more repression set in. Business was shaken by recurrent commercial crises – 1797, 1816, 1825–1826 and thereafter at roughly ten-year intervals until 1866. Not for nothing, however, is John Bull always shown in Regency clothes. He was never more self-confident than in those years, especially after the defeat of Napoleon in 1815. Emerging industrial capitalism was in its Homeric age: rough, hard, pitiless; full of heroic figures, valiant achievement, memorable strife : a prolific begetter of myth and legend still potent a hundred and fifty years later. This vigorous society stood urgently in need of efficient transport for goods and for people. Before the railways, the most practical route for goods in bulk was by water, coastwise or inland. Hence the rapid building, as the pace of development quickened in the latter part of the 18th century, of a system of canals connecting with rivers, each other, and the sea. By the time McAdam and his sons were active London, Bristol, Birmingham, Liverpool and Hull were all linked, besides a great many other towns, and over 4,000 miles of waterway were open for inland navigation. Many unnecessary canals were built, but the general effect of canal building was enormously stimulating, both by opening up new markets for existing businesses and by creating new opportunities as cheap bulk transport became available.[7]

More and more people needed to move themselves about the country as urgently as they needed to move more and more goods, but the improvement of the roads by no means kept up with the building of canals. In fact, it lagged behind by something like fifty years. Why this should have been so is far from clear, but somewhere near the centre of the mystery is the curious fact that from the breakdown of Roman rule until the early 19th century no one in England seems to have considered that the layout, construction and maintenance of roads required any specialized skill. Roads after the Romans, as is still very obvious, were not systematically laid out but brought haphazardly into existence over centuries. As for surfacing and repair, it was generally considered that if cartloads of biggish stones were dumped at intervals the traffic would work them into some sort of carriageway. It didn't. 'Any person', said J L McAdam in 1811, 'who will take the trouble to follow a carriage on the road, will see that the wheel does not pass over the materials of which the road is formed, but is constantly almost at every step encountering the pressure of the wheel, or the carriage must be lifted up by the force of the cattle* so as to surmount it; the effect on the road is either to leave a hole,

* The usual word for draft horses in the early 19th century.

in case the stone is displaced, or that the wheel makes a hole by falling from the height to which it was forced by the obstacle.' Thus our ancestors – such of them as could afford the expense of a carriage, post-chaise or stage-coach – bounced and crashed about their business or their pleasure. The effects of the process McAdam describes might be merely uncomfortable or positively lethal. On the evening of 30th November 1812 a coach near Bath hit 'a heap of unbroken Stones in the middle of the Road' and was upset. Several of the passengers were injured and the coachman was killed.[8]

Behind this deplorable situation lay a system of administration, or frequently non-administration, based on parishes and turnpike trusts. These were local bodies, as local as could be, not even as large as the county which was commonly taken as an administrative unit. The idea that the upkeep of the King's highway was a local, not a national, responsibility ran high into antiquity, and was expressed in a considerable body of law which in Regency England was mainly governed by two statutes of 1773 : the General Highway Act and the General Turnpike Act.[9]

The General Highway Act, following older statutes and even older common law, imposed a duty on the inhabitants of parishes to keep highways repaired by their own unpaid labour or the labour of horses drawing carts for a specified number of days in the year. No one, naturally, expected anyone except the poor to turn out for labour in person, and the Act legalized an already accepted practice by which those who preferred to pay rather than work were allowed to do so. To assemble and direct statute labour, to collect money from those who compounded, and generally to see that the roads were properly looked after, the Act, again with an eye to earlier laws, provided for the yearly appointment of ten 'surveyors' in each parish. Property qualifications were laid down, but no evidence of technical competence was required. The surveyors were unpaid. Their duties, if they carried them out, were well calculated to make them unpopular. If they did not, the roads would be neglected and they might be punished.

The Roads of the Regency

The law thus provided, in theory, for labour, money and supervision for the care of the roads. In practice the labour was unwilling and unskilled, the money was scarce, and the supervision, undertaken as unwillingly as the labour, was unqualified. It was a system barely adequate for looking after a few rough country tracks, but it was the only system publicly provided for keeping up the main roads which served the expanding economy of Regency England.

Alongside it there was another system, linked to it only by a right to call for statute labour or money instead. As far back as 1663 the deficiencies of public provision for road maintenance had led to an experiment with private enterprise, and the first turnpike* gate in England had been set up at Wadesmill on the Cambridge road in Hertfordshire. During the 18th century, especially the latter part of it, roads were 'turnpiked', to use the contemporary jargon, ever more widely and increasingly rapidly, particularly along the most important routes. By 1818, when McAdam was reaching the height of his influence, 19,275 miles of road, about 17 per cent of the total mileage of road in the country – 114,379 – was in the care of turnpike trusts, and the turnpike mileage was still growing rapidly. By 1823 it reached 24,599 miles. The railway route mileage at its greatest extent, 103 years later, was 20,405.[10]

Turnpike trusts in Regency England thus controlled a very large mileage which included virtually all the main roads and many by-roads as well. It was through their agency that the great improvements to roads, with which J L McAdam's name is linked, were effected. It was for turnpike trusts and, in Sir James McAdam's case, that very similar body the Commissioners of Metropolis Roads, that McAdam and his sons worked. It is time we looked at turnpike trusts more closely.

The turnpike principle, still widely acted upon in the United States and elsewhere, is that the upkeep of roads should be a charge on those who travel along them rather than on the community at large. The fully developed English system of the

* The origin of the word, in the late middle ages, seems to have been a spiked barrier placed across a road for defence, especially against horsemen. (*Shorter Oxford English Dictionary.*)

late 18th and early 19th centuries, under which the McAdams did their work, required an Act of Parliament to create a trust responsible for the repair and improvement of a carefully defined road or network of roads : sometimes a 'great road' with branches; sometimes, as at Truro, roads radiating from a town. Usually, a trust's roads were in existence when it was set up. It was rare to establish a trust to build a new line, and if a new line was proposed by an existing trust, a new Act was normally needed before it could be built. A trust's mileage might vary between three or four miles and the 178 of the Bristol Trust, the largest in the kingdom, or 150 at Exeter. In the opinion of Sir James McAdam, founded on vast experience in these matters, there was a simple limit on the mileage which a trust might cover. 'You must limit the extent', he said in 1839, 'by the probability of the attendance of gentlemen at meetings, who will not ride further from home than they can return in an evening.'[11] Some such thought must have been in the minds of the Bath trustees when in 1824 they fixed their monthly meetings for 'the Saturday nearest the Full Moon'.[12] Sir James thought his rule would limit the area of any trust to about twenty miles radius from a centre, but in practice most trusts' areas were much smaller. When the system was at its height, in the 1820s and early 1830s, there were over 1,100 trusts for about 25,000 miles of road.[13]

To put the roads in order in the first place and thereafter for improvements, a trust was authorized to raise capital, usually referred to as 'bonded debt', secured on the tolls. The tolls were for the service of the debt, including a sinking fund to pay it off, and for the expenses of upkeep. It was intended, nominally, that the debt should be paid off during the life of a trust and the roads, improved and in good order, should be handed back to the parishes, but that never happened. In a well-run trust the debt would be regularly serviced and the roads would be well maintained, but invariably when the trust's act ran out there was debt still owing – perhaps increased by improvements – and a new act would be required.

The mortgage bonds of a well-run trust provided an attractive investment, readily realizable and yielding perhaps 4 or 5 per cent. The gentry, whose goodwill was essential for any enterprise in rural England, would be asked to join in the original 'subscription', but after that, if the case of the Bath Trust investigated by Mrs Buchanan is typical, the regular investors would be among the clergy of the neighbourhood and among the surgeons, attorneys, merchants, manufacturers and traders of the country towns, and their womenfolk.[14] It was wise to discriminate between one trust and another. William McAdam, fobbed off by a number of trusts with bonds instead of salary, valued them at 65 per cent of par.

Each turnpike act, besides authorizing a trust to borrow money and charge tolls, provided for the appointment of trustees and gave them powers, such as power to go upon private land in search of road stone, which they would need for managing their roads. They were also given a claim on the statute labour, or 'composition money', of the parishes which their roads traversed, and this could be an important source of income.

By way of executive officers, each trust appointed a treasurer, often a local banker; a clerk, almost always a local attorney; and at least one surveyor. In the 18th century the clerk and the treasurer were often the same, but the General Turnpike Act of 1822 put a stop to that. The clerk and the surveyor were both paid, but each would need other sources of income.

Of the surveyor, we shall have much more to say (see pp. 45–8). As for the clerk, his salary would hardly be the main attraction of the job. £100 a year, voted in 1810 to Philip George, clerk of the Bath Trust, was a high figure, but scarcely a major inducement to a prosperous attorney.[15] A large and active trust, however, and Bath was both, had frequent need of an attorney's professional services. There would be counsel's opinion to be sought; litigation to be threatened, launched, or resisted; negotiations with property owners and other trusts to be conducted. For all these, and more, Mr George had a right to his fees. So had his colleagues serving other trusts, but few were so profitable to their clerks as Bath.

There were fees, above all, when a trust's act fell due for renewal. The fiction that the trust would be wound up after twenty-one years had long been abandoned by the early 19th century, and renewal was invariably sought. New powers might be needed, there might be opposition, in any case the bill would have to be seen through parliament. All this was expensive. In 1810 Philip George presented his trustees with a claim for £1,184 7s 8d in connection with getting a new act, and when the trustees were again seeking a new act in 1829, four of them were deputed 'to attend Mr George in the prosecution of the Bill through . . . Parliament', and they were empowered to require the assistance of the surveyor as well. For the purposes of the bill, Mr George was granted £500 in March, £500 more in May, and in June the minutes record that 'vexatious opposition' to some parts of the bill would greatly raise the cost of getting the new Act.[16] For a small trust the expense of getting a new act might be ruinous, yet in spite of agitation it did not become unnecessary until 1831 when the first Annual Continuation Act made renewal automatic unless there were objections.

Setting up a turnpike trust, financing it, running it, renewing it, were matters for local initiative arising from local requirements. Any wider interest was secondary, unless it was thrust under trustees' noses as those at Staines had it thrust under theirs when the Post Office authorities indicted them for failing to keep their road through Egham fit for His Majesty's mails.[17] On a journey of any distance – London to Bath, say – the traveller passed frequently from the area of one trust into the area of another, sometimes with startling differences in the quality of the road.

The road from London to Holyhead was placed in 1815 under Parliamentary Commissioners with wide powers, which they used with vigour, and in 1827 fourteen trusts in London, administering 131 miles of road North of the Thames, 'were consolidated' under another body of Commissioners (see Chapter VII), but in general the unit of highway administration in early 19th century England was the turnpike trust –

small, local, fiercely autonomous. In that there was nothing peculiar. Until comparatively recently – to a large degree, right down to 1914 – most government was local government insofar as it affected daily life. Westminster and Whitehall did not presume to interfere.

This was in many ways an admirable principle, but when applied to highways it meant there could be no central planning of trunk routes and that roads of national importance had to be kept up out of local resources, often inadequate. A trust set up for local convenience might be driven into bankruptcy by debt incurred to try to meet the needs of long-distance traffic: a manifest injustice and a fertile source of complaint. The trustees of two trusts – St Albans and Whetstone – to which James McAdam was surveyor in the 1820s, quarrelled for years with the Holyhead Road Commissioners on this very point (see Chapter VII), and in 1833 the trustees of the Old North Road between Royston and Wansford Bridge, which carried the Edinburgh mails, appealed for help to the Treasury. They did not get it.[18] A large turnpike trust, such as that at Bath, might have over a hundred trustees. Anything to do with roads, then as now, might be contentious and for the same reason: rights of ownership and property might be affected. It was desirable to make as sure as possible of influential support in the neighbourhood and so the compliment of a trusteeship would be offered, chiefly to landowners, fairly widely, and in a crisis many trustees might attend a meeting and vote. One hundred and nine trustees attended one such meeting at Bath in 1817.[19]

In the ordinary way, the running of the affairs of a trust would be undertaken by such of the trustees as were willing to give their time to it: a group, perhaps, of a dozen or so, perhaps fewer. If they were conscientious the work – unpaid, of course – might be quite heavy, involving ordinary general meetings, perhaps as often as once a month, though more often quarterly, occasional special meetings, and committee work involving travel about the district to view the roads and the composition of reports on problems as they arose, such as

those which in 1820 caused the St Albans trustees 'no inconsiderable degree of Alarm' and led to the appointment of
James McAdam as surveyor.[20] Not all trustees were conscientious, but during the 1820s and 1830s, at least, there seems
to have been no great difficulty in finding sufficient who were.

Trustees, naturally, were drawn exclusively from the fairly
small number of individuals who might be expected to be
influential in local or national politics. At the head of the list
would be any great noblemen who had estates in a trust's
district. Thus the Marquess of Salisbury was a trustee of
several trusts in and about Hatfield, and the Earl of Hardwicke,
similarly, was a trustee in Cambridgeshire, where part of the
Great North Road ran close to his seat at Nuneham Wimpole,
and in Hertfordshire, where he had another house, Pierrepoint,
on the Holyhead road.[21] The Duke of Richmond, whose
principal English seat was at Goodwood in Sussex, took an
interest as a racehorse owner in the road across Newmarket
Heath, and James McAdam in 1833 took care to lay before
him 'the plan of a little Improvement I am anxious to effect',
though whether the Duke was actually a trustee does not
appear.[22] The Sparrows Herne Trust, which looked after 26
miles of the road which is now A41 from Bushey Heath almost
to Aylesbury, had in the early 1820s three earls among its
trustees – Bridgewater, Essex and Clarendon – and they were
fairly assiduous; Clarendon, a very poor earl with only £3,000
a year, but distinguished ancestry, being for some time
Treasurer.[23]

The country gentry were numerous among turnpike trustees,
and among the active trustees some of the most conspicuous
are the clergy, attending regularly at trustees' meetings, sometimes in groups of five or six, occasionally acting as chairman
or treasurer, taking a fair share of committee work. Then there
was an equally large class of trustees referred to in the minutes
as 'gentlemen' but not allowed the rank of Esquire in the
attendance lists, being written down simply as Mr. Presumably
they were substantial townsmen or tenant farmers, without
landed estates. Among the trustees of Wadesmill, in the 1820s,

were three called Fordham, apparently brothers, but only one was Esquire. They were presumably of the banking family, Fordhams of Royston, who were bankers to more than one trust in that neighbourhood. A banker among the trustees seems to have been desirable, and famous banking names – Baring, Hanbury, Thornton, Archer Houblon – appear in the minutes of trusts in the localities where they lived.

Turnpike trusts, as will by now be evident, were by the early 19th century part of the close-knit, stable texture of pre-industrial county society and county administration. They were linked with Quarter Sessions by trustees who were magistrates, with Parliament by trustees who were peers or MPs, or could bring parliamentary influence to bear by family or social connections. They were linked to each other by plural member-ship, which must have considerably increased the work done by the more active trustees and at the same time provided more continuity of administration between adjoining sections of road under the control of different trusts than would otherwise have been possible.

The trustees represented the land-owning and farming interests in the countryside; the trading, professional and banking interests in the market towns. They usually met in one of the posting houses which relied on the coaching trade for their living – the King's Arms at Berkhamsted, the Hardwicke Arms at Arrington, the Feathers at Wadesmill, to mention a few which still exist – or in public buildings like the Guildhall at Bath and the Town Hall at St Albans. They provided lucrative practice for generations of country attorneys – Georges at Bath, Grovers at Hemel Hempstead, Motts at Much Hadham – some of whose firms still exist. The Motts held the Wadesmill clerkship from 1792 until the end of the trust in 1872.

Lowest in the hierarchy of trust officials, exposed to all the hostility which the public felt towards the turnpike system, came the gatekeepers. With a house provided and wages which, at Truro in 1828, ranged upwards according to the importance of the gate from 10s to £1 a week, the job had its attractions in the poverty-stricken post-war countryside, in spite of the twenty-four-hour attention to duty which it demanded.[24] If popular anger was aroused against the turnpikes, as it was in North Somerset at various times in the 18th century and in Wales in the 1840s, gatekeepers might be in physical danger, but in normal times they were looked on much as traffic wardens are today: the traveller's natural enemies, to be regarded with a grudging tolerance and cheated whenever possible.

There are occasional reports in trustees' minutes of evasion of tolls and assaults on gatekeepers, and in 1822, near Tring, a gatekeeper and his wife were murdered, probably for the sake of the money in their till. By the 'prompt exertions' of a clergyman and other inhabitants of Tring the two murderers were almost immediately caught. The trustees, led by two Earls – Bridgewater and Clarendon – decided to pay £12 10s for the funeral of the couple, 'old Servants of the Trust', for the expense of the prosecution, and for a reward of £40 to be distributed among eight men, one of whom, apparently finding blood money repugnant, declined his share. The trustees 'expressed their approbation of Atty's handsome conduct'.[25]

Manual work on the roads was considered the lowest form of unskilled labour, fit only for paupers, old men, women and children or for farm hands who would otherwise be unemployed. Of these there were many, and the McAdams took care to point out, on all suitable occasions, that their system provided work for the poor.

Wages, which are discussed in more detail later (see pp. 54–5), were linked to the agricultural wages of the district, but the McAdams, whenever they could, put their men on piece-work. At 10d a ton, said J L McAdam in 1823, workmen were 'very desirous of contracts' for breaking stone 'because the heavy work is done by the men, the light work with small hammers by the wives and children, so that whole families are employed.'[26]

Many trusts let their tolls at auction, which provided a guaranteed level of income (until the railways began to upset things) and relieved the trustees of direct supervision of the gatekeepers, who were employed by the toll-farmers. Some toll-farmers, particularly round London, operated on a large scale and in co-operation with each other. In the records of trusts in Hertfordshire and Essex which employed James McAdam several names recur, sometimes as farmers of the tolls, sometimes as sureties for each other. Two are especially prominent: Elisha Ambler and Lewis Levy. Levy is said to have 'farmed from £400,000 to £500,000 of turnpike tolls within a radius of sixty to eighty miles from London'[27] and he was in business for many years.

Until the railways began to spread in the late 1830s traffic on turnpike roads was varied, heavy and increasing. Coaching directories make it clear that complicated journeys were possible and no town of any size was likely to be inaccessible by coach from any other. Cary's directory of 1828 lists nearly 600 principal services based on London, a very large number 'from London to the circumjacent villages', and 983 services based on provincial towns. Between these various lists, and within them, there is a considerable overlap, but chance references in turnpike trustees' minutes reinforce the impression of busy coaching traffic. In 1826, for instance, not less than twenty coaches passed daily through Watford,[28] and important coaching centres, as will appear later (see p. 43), were much busier.

For those who wanted privacy or speed, or both, post-chaises could be hired, but the charges were high. *Paterson's Roads* for 1826 quotes rates between 12d (5p) and 18d (7½p) a mile for a pair of horses, adding 'two pair of horses are charged usually at double, and a single horse at half the price of a pair', and *Cary's New Itinerary* of 1828 carries the rates up to 2s 6d (12½p). 'If . . .', Dr Johnson is alleged to have said, 'I had no duties, and no reference to futurity, I would spend my life driving briskly in a post-chaise with a pretty woman.'[29] Delightful, no doubt, but expensive.

Private carriages, even more, were for the rich. By the time the McAdams were active there was a great variety of them, particularly of the light, owner-driven kind, the sports cars of their day. The Bath Trust, in 1829, proposed to charge a toll of 4d 'for every Horse or other Beast drawing any Coach,

Barouche, sociable, berlin, chariot, landau, chaise, phaeton, curricle, gig, caravan, cart upon Springs, hearse, litter or other light carriage except Stage Coaches.'[30]

At the other end of the scale were the great lumbering carriers' wagons on very broad wheels drawn by teams of horses (p. 31) which, with their loads, might weigh six or eight tons. Regular goods services by 'fly wagons' and the somewhat lighter 'fly vans' covered the country much as the stage-coach services covered it for passengers, and in some districts there were specialized carrying trades, such as those for coal and stone near Bath (see p. 121). Then there was local traffic, including farm wagons, market gardeners' carts and so on, and droves of animals moving either over short distances or long.

All this traffic caused parking problems. In 1825 James McAdam, as surveyor to the Cheshunt Trust, was told to prevent carriages standing in the road at Hoddesden and to proceed against owners who refused to move them. It is not recorded whether he was empowered to tow them away. At Royston, in 1832, carts and wagons were frequently left standing in a road only 32 feet wide, by the side of a pond with dilapidated railing, while the drivers were in 'the contiguous Public House'. Accidents resulted, including the overturning of one of Lord Hardwicke's carts into the pond, and 'a foot passenger also was precipitated into it from the footpath in the dark'. The trustees thought the pond 'had better be filled up' and suggested forming a parking place nearby, with a wide and convenient footpath, by levelling a mound. Lord Hardwicke, who was one of the trustees, evidently had no mind to see any more of his carts shipwrecked.[31]

Road transport was chiefly in the hands of three inter-connected groups : inn-keepers, coach and wagon proprietors and horse-masters. The scale of their operations in the years of J L McAdam's greatest influence may be judged from evidence given before a Select Committee of the House of Commons in 1819. William Waterhouse of the Swan-with-two-Necks in Lad Lane, one of the greatest of the posting houses,

said he didn't keep the house, he was the coach-master, owning
many mail and other coaches and 400 horses working generally
up to 50 or sometimes 100 miles from London. At the Golden
Cross, Charing Cross, William Horne had 400 horses and forty
coaches which he worked halfway to Bristol. 'With Mr
Pickwick of Bath', he said, 'I work to Newbury'. John Eames
kept the White Horse, Fetter Lane, and owned the Angel,
St Clement's. He had 300 horses. Outside London, too, there
were substantial men. George Botham of the George, Newbury,
had over 100 horses. Edward Fromont of Thatcham, who said
he had fifty years' experience, was working different coaches a
total distance of 500 miles a day on the Bath Road.[32] William
Chaplin (1787–1859), the greatest coach-master of all, began
life as a coachman, switched into railways in the late 1830s,
became Chairman of the London and South Western, and died
worth half-a-million.[33] Running coaches was not a gentlemanly
occupation. Coach-masters who gave evidence to the Select
Committee were prefixed 'Mr'. None was dignified with 'Esq'.

Gradients were steep: James McAdam considered 1 in $22\frac{1}{2}$
'an easy trotting slope',[34] 1 in 12 was common, Kingsdown Hill
on the London road near Bath had 1 in 9. Loads were heavy: a
laden coach might weigh two tons, to say nothing of carriers'
wagons. Coaching speeds were high: sometimes ten miles an
hour or more. Roads, especially round London, might have
what coachmen called a 'heavy surface' making traction difficult.

The result was spectacular cruelty and the callousness which
the custom of the trade induced. 'My horses', said John Eames,
'upon an average, don't last above three years in the fast
coaches'. Waterhouse calculated that his stock, working up to
50 miles out of London, would not last more than four years.
Outside London, matters might be rather better. 'First', said
William Horne, 'the work is lighter, and next, the food is
better; besides which . . . the stables are airy and more healthy;
they have not so often diseases in the country as we have in
London'. The average working life of a coach-horse in the
country, these London coach-masters thought, was about six
years.

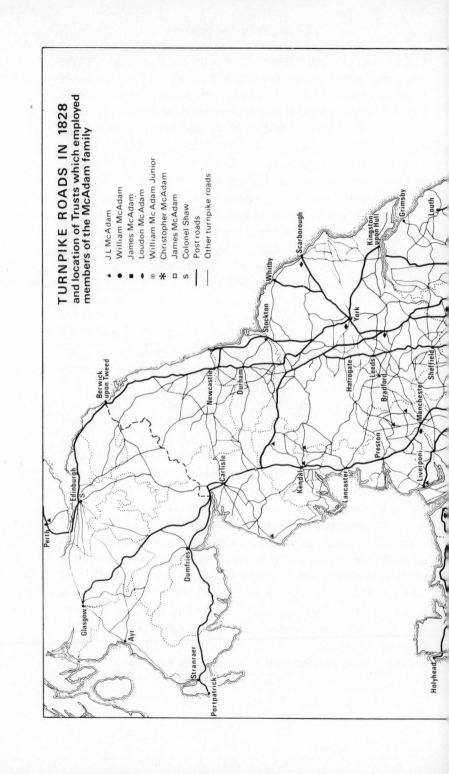

TURNPIKE ROADS IN 1828
and location of Trusts which employed
members of the McAdam family

◄ J L McAdam
● William McAdam
■ James McAdam
◆ Loudon McAdam
⊙ William Mc Adam Junior
✳ Christopher McAdam
□ James McAdam
S Colonel Shaw
— Post roads
— Other turnpike roads

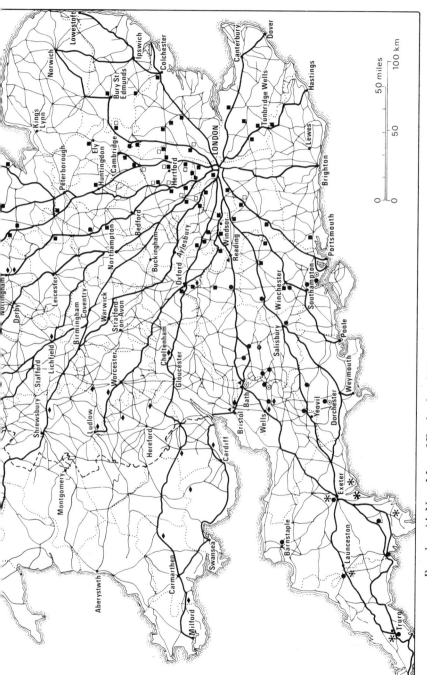

Based on 'A New Map of England and Wales, adapted to Cary's New Itinerary', 1828.

Macadam

Romance gathered about stage coaches as soon as they disappeared. In their day there was little enough. For passengers and crew they were cramped and cold – there are tales of outside passengers freezing to death – and they had not even the merit of being cheap. For horses they were engines of hell. Stage coaches, like the steam locomotives which triumphantly displaced them, and like sailing ships, look much better from a distance than they ever did close at hand.

Even in their inadequate condition at the beginning of the 19th century, the roads of England had been greatly improved by the turnpike trusts, as the rapid increase in the speed of travel showed. To take but one example, in 1754 the journey from London to Bristol took two days, but thirty years later some coaches were doing it in sixteen hours.[35] They were not good enough, however, and in the early part of the new century parliamentary committees enquired into them frequently and heard evidence at length. Great noblemen like the Marquess of Salisbury and the Earl of Hardwicke took the matter up and so did politicians like Sir John Sinclair (1754–1835), twice President of the Board of Agriculture (1793–1798; 1806–1813); the Earl of Chichester (1756–1826), Postmaster-General (1807–1826); Sir Henry Parnell (1776–1842), Chairman of the Holyhead Road Commissioners, a strong supporter of Telford and an enemy of the McAdams. Telford himself, unlike other civil engineers, applied himself to the despised art of road-making.

For those with an ambition to improve English roads, the only administrative framework ready to hand, apart from the antique buffoonery which went on in the parishes, was that provided by the turnpike trusts. On turnpike trusts, by 1816, the leading authority in the country was J L McAdam, Esq., of Bristol.

The Cambridge Telegraph outside a large posting house in Fetter Lane, London. James McAdam was surveyor to most of the trusts on the road between London and Cambridge (p. 79)

Fotomas Index

The yard of the Blossom Inn, Lawrence Lane, London, famous for three centuries as a carriers' inn. Bedrooms open on to galleries, rather like a motel.

Fotomas Index

An Eighteenth Century Road near Highgate, London (pp. 5–6).

Museum of London

REFERENCES

1. W T Jackman, *The Development of Modern Transportation in England*, 2 vols CUP 1916, I 339.
2. J L McAdam, *Observations on the Highways of the Kingdom*, printed as Appendix C, No 1, to the Report from the Committee on Highways and Turnpike Roads 1810–11, 29, and as Appendix to J L McAdam, *Remarks on the present System of Road Making*, Longmans 1823, 34–5; Minutes of Evidence to the Select Committee on Highways 1819, printed in *Remarks*, 113; and see *Observations* and *Remarks* generally, also McAdam's evidence to the SC of 1819.
3. McAdam's evidence to SC of 1819, reprinted in *Remarks*, 97.
4. McAdam's evidence to the SC on Mr McAdam's Petition, 1823, Minutes of Evidence 11–12.
5. Peter Mathias, *The First Industrial Nation*, Methuen 1969, 136 (paperback edn).
6. F M L Thompson, *English Landed Society in the Nineteenth Century*, Routledge & Kegan Paul 1963, 25.
7. As (5) 111–112.
8. *Observations* as (2) 29. Bath Trust Minutes 5xii1812, Somerset RO.
9. As (1) 218–227.
10. SC 1820, Report, 5; SC 1823 (as (2)), Report 12–13; B R Mitchell and Phyllis Deane, *Abstract of British Historical Statistics*, CUP 1962, 227.
11. SC 1839 Minutes of Evidence 472/39.
12. Bath Trust Minutes 20xi24.
13. Sir James McAdam's figure for the number of trusts – Abstract of the General Statement of the Income and Expenditure of Turnpike Trusts in England and Wales 1834, Goodwood MSS 1874/221.
14. B J Buchanan, *Capital Formation in North Somerset 1750–1830*, PhD Thesis Bristol 1979, 367–447.
15. Bath Trust Minutes 11vi10.
16. As (15) 28ii 21iii 20vi, 1829.
17. SC 1823, Appendix B 80.
18. Kneesworth & Caxton Trust (Old North Road) Minutes viii1833, Cambridgeshire RO.
19. As (15) 15iii17. SC 1823 Minutes of Evidence 59 gives the number present as 107.
20. St Albans (Pondyards & Barnet) Trust Minutes 4i20, Herts RO.
21. Cary's *New Itinerary* 1828, 446 290.

Macadam

22. Sir James McAdam to the Duke of Richmond 9i35, Goodwood MSS 1463/363.
23. Sparrow's Herne Trust Minutes, Herts RO; as (6).
24. Truro Trust Minutes 25vi28 Cornwall RO.
25. As (23) 22xi 6xii 20xii 1822, 8i23.
26. Sir James McAdam to Wm C Kitchener, as (23) 13ix46.
27. Edwin A Pratt, *A History of Inland Transport and Communication*, 1912 (David & Charles Reprints 1970), 318.
28. As (21), no page no; as (23) 17iv26.
29. Boswell's *Life of Johnson*, 20ix1777.
30. As (12) 24iv29.
31. Cheshunt Trust Minutes 9viii25, Herts RO; as (18) 26iv32.
32. SC 1819.
33. N W Webster, *The Great North Road*, 1974, 20–21.
34. Sir James McAdam to the Duke of Richmond 9ii33, Goodwood MSS 1463/429.
35. As (1) 293.

22

II

The Making of a Road Surveyor

Two False Starts and a Failure, 1756–1801

JOHN LOUDON MCADAM was born in Ayr on 21st September 1756, the youngest of ten children, eight of them daughters, of James McAdam of Waterhead and Susannah Cochrane, niece of the 7th Earl of Dundonald. The marriage, surely an ambitious match for a minor laird, is said to have been arranged by James's guardian (his father was dead) when James was seventeen and his bride was a year older. John Loudon's parentage conferred on him the standing of a gentleman which he greatly valued, in later life, for the lustre which it shed on his rather humble occupation of road surveyor. In documents where his name appears he is always Esquire rather than Mr: a distinction by no means automatically awarded to professional men and never to men in trade.

James gradually sold off the Waterhead estate, it is not clear for what reason, and in 1769 he put £500 into the Ayrshire Bank, largely financed by landowners, which failed in 1772, but when he died in 1770 he was far from being a ruined man. Indeed, if a niece of his wife is to be believed, he was tolerably well off, especially by the Scottish standards of the day, for she says 'he died worth about 6 or 7000 £'. He had numerous daughters to provide for. Miss Cochrane says he left them £500 each, and the residue went to his only surviving son, Loudon, 'a promising boy yet at school'. It seems likely, therefore, that Loudon had at least £1,000 or so to console him for his father's death, and possibly a good deal more.[1] He was an orphan, but never poor.

He had nevertheless his way to make, and in a characteristically Scottish manner he went abroad, to the colony of New York. There his uncle William McAdam (1725–1779) was a prominent merchant. He had gone young to America, married a well-connected wife, Ann Dey, who was ten years younger than himself and survived him by forty-nine years, and established himself firmly in New York society. In 1766 he had a warehouse on Smith Street, near the New Dutch Church, where he advertised 'Iron-bound Butts and Puncheons, genuine Batavia Arrack in Bottles, Frontinjack, Priniack and Madeira, etc.' In April 1768 he helped to found the New York Chamber of Commerce, of which he became Treasurer in 1774* and Vice-President in 1775. He remained consciously Scottish, being 11th President of the St Andrew's Society of New York in 1772-1773, and he kept in touch with his Scottish relations. There is indeed a note in the records kept by the New York Historical Society which says that his brother James, father of J L McAdam, followed him to New York about 1762 and 'embarked in mercantile speculations'.[2] If he did, they were probably unsuccessful, for he died in Scotland, but the episode might explain William's readiness to welcome his nephew in 1770: the more so since he lost his own children and might by that time already have been childless and looking for someone to keep his business in the family. J L McAdam's obituary notice in the *Gentleman's Magazine* of January 1837 says that his uncle adopted him.

* Mrs Pember says, I think wrongly, that J L McAdam, not his uncle, became Treasurer in 1774, being only eighteen at the time.

The Making of a Road Surveyor

J L McAdam, according to records of the St Andrew's Society in New York, 'passed his apprenticeship in some mercantile establishment' which may or may not have been his uncle's. He seems, in any case, to have passed his youth in circumstances of some affluence, for in the spring of 1776 he made a tour of the West Indies, perhaps partly on business but not entirely so. In Bermuda he spent a week with Mr and Mrs Tucker. Tucker was a member of the Legislative Council and his wife, whose maiden name was Bruyère, was a daughter of the Governor. She might have married McAdam's brother, years earlier, if the young man's father had not disapproved. She bore J L McAdam no ill-will but her 'old scold of a mother', he wrote in 1832, attacked him whenever he went to Government House.[3]

By the time McAdam came back from his tour in mid-August 1776, fighting between the forces of the Crown and the American colonists, which had broken out in the previous year, had reached New York. For a time McAdam could not get home because a fleet and troops under the brothers Howe were preparing to retake the city from the rebels, which they did at the end of the month. Uncle William at that time had 'retired to the back part of New Jersey', hoping to keep out of harm's way.[4]

No doubt many people shared William McAdam's un-heroism. He had gone so far in May 1774 as to be one of the New York Committee of Correspondence's fifty-one gentlemen 'to correspond with the neighboring colonies', but he was not prepared to go into rebellion, and there were many like him, particularly in New York. The Atlantic was wide but connections across it were close, multifarious and highly personal, and the thought of war between Englishmen and against the Crown was deeply shocking. The younger McAdam made his position clear at the outset and never budged from it, referring to the rebels with what looks like distaste as 'Americans'. By December 1776 at the latest, possibly earlier, he was in business for himself, and with the permission of Major-General James Robertson (1720?–1788), who had risen from the ranks and was to become, in 1781, civil Governor of New York, he advertised a sale of prize goods at his store on New Dock.[5]

In March 1778 he felt secure enough to marry. 'Married a few days since', says a report of March 1778, 'John McAdam of this City, Merchant, to Miss Glorianna Margaretta Nicoll, Daughter of William Nicoll Esq. of Suffolk County, on Long Island, a young Lady of great Beauty and Merit with a large Fortune.'[6] Marriage cemented still more firmly the alliance between McAdam and the solid Establishment of colonial New York. His bride's father, a well-to-do lawyer, presided over the Provincial Assembly when it declared New York independent, but he showed no enthusiasm for pursuing the matter further although regarded with some suspicion by the Crown forces in Long Island. The Nicolls were related to the De Lanceys, another old and distinguished New York family who were prominent loyalists. Brigadier-General Oliver De Lancey (1749–1822) commanded a brigade of loyalists, raised and equipped at his own expense for the defence of Long Island, and J L McAdam served in it, presumably part-time. The first battalion went to Georgia in November 1778, but the other two battalions never left Long Island, so it is unlikely that McAdam's service was arduous, since Long Island was never attacked. De Lancey, a regular cavalry officer, had a long career in the British army after the Revolution, varied by a spell as MP for Maidstone from 1796 to 1802 and jolted by being removed from his position as Barrackmaster-General, in 1804, for culpable carelessness in keeping his accounts : a misfortune which did not interfere with his promotion, for he became a full General in 1812. The De Lanceys were 'extensively intermarried with prominent English families', according to an American genealogist of 1848, and held 'high and honorable appointments in the government of Great Britain'.[7]

Deeply committed as McAdam was by sentiment and marriage to the loyalist cause, there would have been no place for him in New York after peace was made, any more than there would have been for his uncle, posthumously attainted and proscribed by the rebels on 22nd October 1779.[8] A new partnership, McAdam Watson & Co., 191 Queen Street, New

York, was formed in April 1781 but in November 1783 it was dissolved.[9] By that time McAdam was in Scotland with his wife and his two eldest children, Anne and William, for they arrived there from America in June 1783.[10]

McAdam's situation was by no means so forlorn as that of loyalists who found themselves compelled to put their lives together again in the inhospitable forests of Eastern Canada. He was able to pay £3,000 for Sauchrie House, between Ayr and Maybole, where he set up in some style, becoming Deputy Lieutenant of Ayrshire and in 1794, when there seemed to be danger of a French invasion, being given command of 'a corps of volunteer artillery'.[11] For about fifteen years, from 1783 to 1798, McAdam was a country gentleman, managing his estate and attending to public business, including obligatory service, beginning in 1787, as a trustee of the Ayrshire turnpike roads.

During this period McAdam also went into business in association with his brilliant, eccentric kinsman Archibald Cochrane, 9th Earl of Dundonald (1749–1831), who was in the early stages of ruining himself with ingenious ventures into the chemical industry. McAdam was nearly ruined, too, and in later life his references to this phase of his career, in evidence to a Select Committee in 1823, were evasive, even downright false. The truth remained hidden until Dr Robert H Spiro searched it out in the late 1940s.[12]

One of Dundonald's enterprises was the British Tar Company, set up at Muirkirk to work a patented process for distilling tar from coal, chiefly with ships' bottoms in mind. McAdam in 1823 said he never managed the works, but Dr Spiro has shown that he became plant manager in the mid-1780s. In 1790 he borrowed £14,000 and bought the business. Then his troubles began. The British Tar Company had a contract with the neighbouring Muirkirk Iron Company for the supply of coke. It was badly drawn, and years of expensive quarrelling followed, dragging on until 1803. After that the tar company supplied McAdam with an income for about a quarter of a century, but the story very nearly had no such happy ending.

By 1795 McAdam owed £14,100 to Admiral Keith Stewart. In that year the Admiral died, leaving McAdam hopelessly in debt to his estate, of which McAdam was a trustee. He was very close to bankruptcy, but it appears that the estate's creditors chose to be lenient. McAdam raised £8,500 by the sale of the Sauchrie estate and left Scotland.

Quite how McAdam spent the next three years, which can hardly have been the happiest in his life, nobody knows, though it is fair to infer that he found some means of repairing his finances. He was himself extremely reticent, giving the misleading impression to the Select Committee of 1823 that he passed the time in Bristol, whereas apparently he travelled to Falmouth in 1798, pausing in Bristol only for his wife to give birth to a third son, John Loudon Jr. At Falmouth, the Dictionary of National Biography says he became a victualling agent for the fleet. That would no doubt have been lucrative, but Dr Spiro found no mention in Admiralty records of McAdam as a victualling agent nor of any victualling agent nearer Falmouth than Plymouth. He did, however, find evidence to suggest that McAdam became a prize-master much as he had done in New York, perhaps with equally gratifying results. He certainly travelled several times to Scotland on the affairs of the British Tar Company. What is of great importance is that according to his own account, which on this point there is no reason to suspect, he began his investigations into the affairs of turnpike roads.[13] No doubt the discomfort of his journeys to Scotland supplied the incentive and their length, the opportunity.

In 1801 the Peace of Amiens put an end to war and prize-mongering. Presumably for that reason, McAdam left Falmouth. He went to Bristol, no one knows why. Why did he not go back to Muirkirk and the tar company which still belonged to him and of which his son William became manager? Perhaps he was too proud to go back, being always a prickly character. When he was famous, he returned.

In 1801 McAdam was forty-five and his life was running unfulfilled into the sand. He had made two false starts, in America and at Falmouth, and one resounding failure. Now, in middle age, he was starting in a new direction altogether, towards the work on roads which would make him famous. He had no formal qualifications, but that was not important. The technical occupations that support industrial society were embryonic, lacking in most cases the body of theoretical knowledge and the code of practice required for 'professions' in the modern sense. What was needed was to define a task and show how to tackle it. This, in the case of road management, McAdam did superbly well, lifting his career out of failure and making of it a brilliant success.

The Inspector of Roads, 1798–1815

From 1798 onward, McAdam told the Select Committee of 1823, he made road inspection 'a sort of business. Without saying to anyone what my object was, I travelled all over the country . . . It was only occasional travelling of course. I had some other occupations and private affairs to look into.' He claimed to have travelled 30,000 miles between 1798 and 1814. Between 1810 and 1813, he said, he had been in very bad health : nevertheless in 1813 he 'visited all the military roads in the Highlands, and many other Scottish roads, as far as Inverness.' McAdam undertook these journeys at his own expense, impoverishing himself in the process. In 1823, when he was seeking a grant of public money to cover his expenses, he explained his methods of gathering information.

He was not content simply to inspect the condition of the roads. He sought out turnpike trustees (whom he usually called 'commissioners') and questioned them about costs. 'I generally found', he said, 'that the expense was in proportion to the badness of the roads . . . and I found few roads that were not deeply in debt and in distress for want of money . . . I believe', he went on darkly, 'there was a great deal of . . . prodigality, of a worse character than carelessness', and on another occasion he estimated that 'mis-application' – arising, he was careful to indicate, far more from ignorance or careless-ness than from fraud – might 'amount to one-eighth of the road

revenue of the kingdom at large, and to a much larger proportion near London.'

He did not stop at questioning trustees. He would talk to
anyone who worked on the roads, making it 'a business to
enquire generally of the surveyors, workmen, and people on
the roads, as to the expense of materials, cartage, day labour.'
He 'inquired very particularly' about statute labour, and 'found
the statute labour, when called for, was sent by the farmer to
the roads, but the people seldom did above half a day's work;
and, though the farmer lost the service of his servants and team,
the public did not get it; it was a heavier oppression on the
country than benefit to the roads.'

McAdam's Scots accent alone must have roused the sus-
picions of early 19th century countrymen, accustomed to regard
anyone from a different parish as a 'foreigner'. His persistent
questioning was most unwelcome. 'With surveyors and other
officers of trusts,' he said, 'I found a jealousy and an unwillingness everywhere to give me information. An unauthorized
individual finds it extremely difficult to procure information
of that sort, and I found it so; a very great unwillingness to
inform.' Naturally. What was this stranger up to? What
business was it of his how they looked after their roads? Was
he trying to get them into trouble?

For McAdam all this travelling over bad roads, all this
questioning of suspicious respondents, was extremely hard work,
and when he went to Parliament for money he wished to leave
MPs in no doubt of it. 'Unless I could explain the infinite
difficulties I had in procuring information,' he told the Select
Committee of 1823, 'and the ignorance with which I set off,
and that I had to be taught as I proceeded, it is impossible to
give an idea of the labour I undertook. Gentlemen might think
I had taken a great deal more labour than was necessary for
the object.'[14]

By 1811, according to McAdam's much later account, he was
in poor health, discouraged by opposition, and inclined to give
up the whole business of roads. Then the President of the Board
of Agriculture, Sir John Sinclair (1754–1835), being Chairman

of a Select Committee, applied to McAdam for information, and that renewed his vigour.[15] With other experts, he gave evidence. He was very confident and his views, printed as an Appendix to the Select Committee's Report under the title *Observations on the Highways of the Kingdom*, are forthright and acidulated. They wore well and McAdam reprinted extracts in his best-known published work : *Remarks on the Present System of Road Making*.

'It would be a useless waste of time', McAdam told the Select Committee, 'to enter into any argument to prove the very defective state of the highways of this kingdom.'[16] Wasting no time, therefore, he went on to say that all Committees of the House on roads, of which there had been many, 'seem to have had principally in view the construction of wheeled carriages, the weights they were to draw, and the breadth and form of their wheels; the nature of the roads on which these carriages were to travel, has not been so minutely attended to.' In other words, Parliament had always addressed itself to the traffic on the roads, never to the condition of the roads themselves.

This was true. 18th century Acts laid down regulations for the construction of carts and wagons and, especially, for the width and form of their wheels. The conventional wisdom was that broad wheels did least damage, so legislation went on that principle, going so far as to favour, for the heaviest wagons, great lumbering rollers, sometimes conical in form, rather than true wheels. Regulations made under an Act of 1765 prescribed, for heavy wagons, wheels constructed and arranged to roll a path sixteen inches wide on each side of the wagon. Turnpike trusts charged differential tolls for wheels of varying breadth. All this exacerbated the unpopularity of the turnpike system and hindered the development of lighter, potentially faster, load-carrying vehicles.[17]

McAdam by 1811 had made up his mind that this view of the problem was wrong. Instead of vehicles being made to suit the roads, the roads should be made to suit the vehicles. He went on to lay down, for the benefit of the Select Committee, the principles underlying the system of road-making which later came to bear his name.

'The observations I have made', he said, 'in a period of twenty-six years on the roads of this kingdom, in which time I have travelled over the greater number in England and Scotland south of the Firth of Forth, and the opportunities I had of making comparisons of the different kinds of materials and modes of their application, have led me to form the following conclusions.

'1st. That the present bad condition of the roads in the kingdom, is owing to the injudicious application of the materials with which they are repaired, and to the defective form of the roads.

'2nd. That the introduction of a better system of making the surface of the roads, and the application of scientific principles, which has hitherto never been thought of, would remedy this evil.

'In illustration of those positions, I beg to observe, that the object to be attained in a good road, as far as regards the surface, is to have it smooth, hard, and so flat as that a carriage may stand quite upright; those objects are not attained by the present system, because no scientific principles are applied; but it is presumed they are perfectly attainable in all parts of the country.

'A road made of small broken stone, without mixture of earth, of the depth of ten inches, will be smooth, hard and durable; this is proved by all experience: it seems only necessary therefore to enquire by what means this desirable object may be attained all over the kingdom.'

Although McAdam said that such a system of road-making as he outlined had 'never been thought of', he nevertheless mentioned two stretches of road where he had seen something very like it. One was in Somerset – 'some miles of road near Bridgewater,* where the stone is well broken, and the road is smooth and good, consequently upheld at a very moderate

* Nowadays Bridgwater.

expense.' The other was in Westmoreland – 'the great road on each side, north and south, from Kendal; I mention this as this road is made the *nearest* to the system I recommend of any I have seen in the kingdom.' There was also a stretch of road near Radstock, probably unknown to McAdam, where one of the Bath trustees, a clergyman, had much the best roads in the trust's area in 1810, because he used 'the very best stone . . . broken to the Size of a large Walnut' and spread about six inches thick.[18] It cost about £968 a mile for Bristol limestone on a surface 15 feet wide and it sounds very much like macadam.

McAdam's system of construction was not in fact arrived at by 'the application of scientific principles', as he claimed, but empirically, by watching what happened as traffic passed along a road and drawing the conclusion that stone must be broken to a size 'in due proportion to the space occupied by the point of contact of a wheel of ordinary dimensions on a smooth level surface. This point of contact will be found to be longitudinally about an inch, and every piece of stone put into a road which exceeds an inch in any of its dimensions is mischievous.' It was 'mischievous' because its size prevented it being consolidated with other small pieces of stone by the wheels of passing traffic into the smooth hard surface required. McAdam observed that the large stones commonly used simply lay loose until wheels either pushed them out of the way or crashed down heavily after bumping over them (see p. 6), in either case making a hole in the surface which very readily lengthened into a rut. Then from time to time the repair men came along with 'another large quantity of the same materials', which they laid on 'the old bad foundation; the expense of this proceeding is very heavy, and the effect is the same as at first, that the road is immediately broken up, and carriages dragged over it at a great and unnecessary waste of cattles labour.'

From this misguided proceeding there was one fortunate result. 'All the old roads in this kingdom', said McAdam, 'contain a sufficient quantity of materials to last for a number of years.' Some, through the mass of stone dumped on them, had been raised 'several feet above the common level'. Therefore any turnpike trustees enlightened enough to give McAdam's methods a trial had plenty of roadstone at their disposal. They had only 'to pick or raise up a sufficient quantity of stone, already in the ground, and to break them; this process will be found much less expensive than the procuring, carting and spreading [of] new materials.' In other words, McAdam was saying, his method was not only the best – it was also the cheapest.

But would many trustees take McAdam's advice? Obviously he doubted it: his opinion of turnpike trustees in 1811 was low. In very many parts of the country, he said, the necessity of breaking stones in road-making had been adopted 'as a good principle', but 'without any good effect to the public' because it had been adopted 'without any fixed idea, directed from reasoning, as to the proper and necessary dimensions to which the stones should be broken'. The size desired was expressed in a 'vague and careless manner' in contracts, and little attention was paid to the manner of fulfilling the contracts. Turnpike trustees, he was implying, were deplorably slack. 'Besides', he went on, suggesting that they were not only slack but stupid, 'it has never occurred to trustees on roads that the materials already in the roads would answer the purpose much better than providing an additional quantity, therefore the public money has constantly been expended in procuring and carting new materials, which have been laid over the old ill-broken stone, to the great detriment of the road, and the waste of the funds: almost every well-frequented road in the kingdom is in want of repair under the present system once in six months; and they are at no times in such a state of repair, as to make them easy for carriages to pass over.'

His methods, he maintained, would work in any part of the country with any material locally available, though naturally some kinds of stone were more suitable than others. He wanted to see a mile of road hacked up and relaid, as an experiment: near London with gravel, in Gloucestershire with limestone, in Staffordshire 'by breaking the pebbles and separating them from the sand', in Kent with broken flint, and in the North with whinstone, 'in every case laying the road flat and even, and making use of the materials already in the road.' The total expense, he thought, of all five experiments, might amount to £2,500. It seems as if McAdam at this time was hoping to get Parliament to give his recommendations the force of law. If the experiments were found to prove that a road of the type tested would be 'more convenient to the public for use' and would be 'more likely to stand without repair, whereby a very considerable saving of expense will be made', then, he said, 'it will be expedient to adopt measures, under the authority of Parliament, to oblige all trustees, commissioners and way-wardens to conform to such regulations in making and repairing roads as shall be laid down; and to provide that no accounts shall be passed and allowed respecting roads, until they have been inspected and reported to be made or repaired agreeable to the law.' Such a proposal was not likely to be welcome to the gentlemen of England, whether they were sitting as turnpike trustees or as Members of Parliament, and it did not prevail. The roads were left in local hands and McAdam, denied coercion, had to try what he could do with persuasion.

The experiments McAdam suggested were never carried out, probably not greatly to his surprise, for in *Additional Observations* subjoined to his main paper he foresaw that they might be 'thought too expensive' or, 'as it will much more likely happen', it would be impossible to overcome the opposition of 'surveyors and all the tribe of road-makers, who will oppose such experiment, *vi et armis.*' To counter that opposition, he suggested getting a report on the expense of keeping up the road near Kendal of which he approved, and comparing it with similar reports on certain roads in Scotland, including the road between Ayr and Glasgow where excellent roadstone was available and the surface was so bad that the mail coach was allowed six hours for 34 miles. Road-making, he remarked in a later paragraph, 'is even worse understood in Scotland than in England'.

In these passages there begins to appear McAdam's contempt for all the professional road-makers and road-repairers of his

day. 'I only propose enquiry', he said disarmingly. 'I propose
to do what never yet has been done, to consider the making the
form and surface of roads scientifically.' He was quite sure he
knew what the result would be : the conclusion he himself had
reached, that large sums of money and large quantities of
perfectly good material were wasted through the carelessness
of trustees and the ignorance of those whom they employed,
especially their surveyors.

'When a road is reported in disrepair,' he said, 'or, in other
words, when the large stones are so displaced, that it is no
longer possible to pass through amongst them, then the
trustees enquire what funds are to be found; and if a large sum
is expended, it is supposed that the road is mended; a contract
is made in terms sufficiently vague to be easily evaded, and it is
left to a *surveyor* to see it executed . . . Surveyors are elected
because they can measure; they might as well be elected
because they could sing; but they are more commonly elected
because they want a situation; and this is the source of much of
the evil we are now complaining of.'

This was a matter he would return to again and again, as,
for instance, in 1823 when he described surveyors as 'always,
I think, almost without exception, very low people, many of
them old servants, ruined tradesmen, people without that kind
of energy and character which I think is absolutely necessary
for such a service.'[19] Along with his ideas on preparing a road
surface, his demand for competent surveyors lay at the heart of
his doctrine. It was not enough for trustees to give directions
for road building or road repair. Someone must understand
them and see that they were carried out. Every trust, that is,
needed efficient management, and there was unlikely to be
efficient management with an inefficient surveyor. So strong
were his feelings that in 1816 McAdam, at the age of sixty,
condescended to become a road surveyor himself.

Breakfast in a coaching inn.

Fotomas Index

The joys (?) of coaching on an unmacadamized road, 1828.

Fotomas Index

JOHN W. PETERS

Design for a posting chariot, 1839. An elegant example of the type of vehicle which could be hired, at great cost, for private travel.

Museum of London

A carrier's wagon of 1820, with the very wide, conical wheels of the period (p. 16).

Fotomas Index

APPENDIX TO CHAPTER II

J L McAdam's instructions for repairing a road, as printed with his 'Observations on the Highways of the Kingdom', in Appendix C to the Report from the Select Committee on Highways and Turnpike Roads, BPP 1810–11 III 855.

Directions for repairing a Road.

NO addition of materials is to be brought to the road, unless in any part it is found that there is not already in it a quantity of clear stone, equal to a foot thick. — Materials in addition.

The stone in the road is to be loosed up to the depth of a foot, and broken so as to pass through a screen or harp of an inch in the opening, by which no stone above an inch in any of its dimensions can be admitted. — Materials to be broken.

The road is then to be laid as flat as possible, if it is not hollow in the middle it is sufficient; the less it is rounded the better; water cannot stand upon a level surface. — Road to be quite flat.

The broken stone is then to be laid evenly on it, but if half or six inches is laid on first, and exposed a short time to the pressure of carriages, and then a second coat of six inches laid on, it has been found advantageous in consolidating the materials. — The broken stone to be laid on; better half at a time.

Carriages, whatever are the construction of the wheels, will make ruts in a new made road, however well the materials may be prepared, or however judiciously applied, therefore a careful person must attend for some time after the road is opened for use, to take in the track that is made by the wheels. — Ruts to be filled up.

The only proper method of breaking stone both for effect and economy, is by persons — Method of breaking stone.

sitting; the stones are to be gathered in small heaps when picked up, and women or men past hard labour, must sit down upon straw mats, and break them so small as to pass through a screen or harp of an inch in the opening.

The method of breaking stones by persons *sitting*, is practised in Westmoreland, and part of Somersetshire near Bridgewater; in those two neighbourhoods they have the best roads, and at the smallest expense.

Tools to be used; picks. Sledge hammers, if the stones are very large. Small hammers.

The tools to be used are strong picks to loosen the stone out of the road; if the stone is of very large size, it must be broken into smaller masses by a sledge hammer, employed by an able man, but probably the stones already in the roads, in most parts of the kingdom, do not much exceed ten or twelve pounds weight; such stone is to be broken by persons sitting, with a hammer about 15 inches in length in the handle, and about an inch broad in the face, calculated as to weight to the strength of a woman's single hand; should any stone, such as whinstone, be found too hard for women or old men to break, stronger men must be employed, but in either case they must be made to sit down: a woman *sitting* will break more limestone for a road than two strong labourers on their feet with long hammers, in a given time.

Wheelbarrows and shovels.

Wheelbarrows and shovels are necessary to distribute the materials.

Iron rake. Dimensions of iron rake.

A rake of iron with short teeth, not to exceed an inch and an half in length; the head ten inches long, is to be employed by a *careful* man, in raking the track crossways when the road is first used; this will fill the tracks at once and keep the road level. Tracks will not occur again after the road has settled, the whole mass will become like one solid smooth surfaced stone.

Every road is to be made of *broken* stone without mixture of earth or any other matter; no large stones to be employed on pretence of *bottomong*, nor any sand, earth, or other matter on pretence of *binding*.

Broken stone alone to be used.

A road made of stone effectually broken, will be a smooth hard even surface, it cannot be affected by wet or by frost, and will therefore be equally good at all seasons of the year. Stone, in some form, is to be found in every part of this island, and therefore every road in the kingdom may be equally good.

REFERENCES

1. 'Roy Devereux' (Mrs D Pember), *John Loudon McAdam*, OUP 1936, provides the only published account of McAdam's career, apart from obituaries and the very unsatisfactory DNB article. It is entirely superseded by Robert H Spiro's unpublished PhD thesis *John Loudon McAdam, Colossus of Roads*, Edinburgh 1950, which is based on Mrs Pember's papers and a great deal more. For the Ayrshire Bank, see William Robertson, *Old Ayrshire Days*, Menzies 1905, 266–273.
2. MS 'Notes on Members, St Andrew's Society of the City of New York' by William M McBean (1852–1924), in the archives of the New York Historical Society, Envelope 137.
3. J L McAdam to his daughter Nancy, quoted in Devereux as (1) 27.
4. J L McAdam to an unidentified correspondent (My dear Girl), quoted Devereux 38–39.
5. As (2) env. 137 646. See also (4). DNB for Robertson.
6. As (2) env. 646.
7. Walter W Spooner (Ed), *Historic Families of America*, New York 1907; Devereux 38; Orderly Book of the Three Battalions of Loyalists, NYHist Soc 1917; Jerome B Holgate, *American Genealogy*, Albany NY 1848, 115.
8. As (2) env. 137, quoting Clinton Papers, Vol 5 159.
9. As (2) env. 646.
10. Spiro 82.

11. SC 1823 24.
12. My account of McAdam and the British Tar Company, and of his life at Falmouth, is based on Spiro 102–129.
13. SC 1823 11.
14. SC 1823 11–24 and generally.
15. John Sinclair, *Memoir of the Life and Works of Sir John Sinclair*.
16. Appendix C to Report of SC 1810–11, 27–32. Subsequent references are to this Appendix unless otherwise indicated.
17. Jackman, *Development of Modern Transportation* I, 216–218; John Copeland, *Roads and their Traffic 1750–1850*, David & Charles 1968, 41–2; Edwin A Pratt, *Transport and Communication 1912* (David & Charles Reprints 1970), 45–50.
18. Bath Trust Minutes 10xi10, Somerset RO.
19. SC 1823 12.

III

Surveyor of the Bristol Roads 1816–1825

'I FORMED A theory in my own mind', said McAdam in 1823.
'This theory I got leave to put into practice by being appointed
to the care of the Bristol roads . . . in January 1816.'[1] This
sounds rather like a management consultant offering to carry
out his own recommendations. McAdam never lacked self-
confidence.

By the time McAdam took up the surveyorship he had been
living in Bristol about fifteen years. During that time, or some
part of it, he had two sons – James and Loudon – and two
daughters – Gloriana Margaretta and Georgina Keith – at
home. William was managing the British Tar Company in
Scotland and James is said, somewhat vaguely, to have been
establishing himself in business. The addresses of the various
houses where the family lived – Park Street; Clifton; Berkeley
Square – suggest comfort: hardly opulence.[2]

How McAdam achieved comfort after the disaster in Scotland
is a mystery. In the will of his American aunt, Ann McAdam,

he is described as a merchant. That is a wide term and there is nothing to narrow it down, least of all in his own guarded references to his affairs. Dr Spiro suggests that after the quarrel between the British Tar Company and the Muirkirk Iron Company was composed he had sufficient income to provide him with 'a large measure of economic security and independence'.[3] Perhaps it was this income which paid for his journeys on the turnpike roads and gave him leisure to undertake them.

He certainly had leisure enough, and a sufficient position in the city, to take some part in civic affairs. He was prominent in a controversy which eventually led to the building of a new gaol, built in 1820, and he stood well enough in the city to be appointed a commissioner (trustee) of the Bristol Turnpike Trust. It was his colleagues in the trust who invited him to become their Surveyor.

Bristol in the early 19th century had lost to Liverpool its ancient pre-eminence among English ports, partly because of what has been called 'a certain complacency and inertia which was a serious handicap in the adjustment to new conditions'.[4] The cotton trade in particular, the main growing point of the early Industrial Revolution, was seized by Liverpool. Nevertheless in McAdam's day Bristol was busy and growing, with a population of 68,000 in 1801 which by 1851 had grown to 160,000, and although Bristol's foreign trade grew more slowly than Liverpool's, yet large businesses which depended on foreign supplies – wine, tobacco, chocolate – flourished there. Moreover with the Severn flowing from the Midlands and the Bristol Channel serving South Wales, and the Somerset coalfield and rich farming areas close by, Bristol was extremely important in home trade even though neither it nor its neighbourhood became industrialized on the same scale, or in the same way, as the new areas of activity in the North or over the water in South Wales.

In McAdam's time the prosperity of the districts immediately inland from Bristol was already of long standing. Bristol itself provided a market for farmers and market gardeners and there were coal mines in Somerset and on the edge of Gloucestershire. A network of turnpike roads came into being, particularly dense in North Somerset, during the 18th century, and in the 19th it provided employment not only for J L McAdam but for his sons William and Loudon and for William's son William (1803–1861) whose activities are discussed in later chapters.

From Bristol stage coaches served the Midlands, South and West Wales and the South-West as far as Plymouth. Cary's *New Itinerary* of 1828 listed eighty services between Bristol and other provincial towns, reaching Birmingham, Manchester and Liverpool in the Midlands and North; Milford Haven in West Wales; Exeter, Plymouth and Devonport in the South-West; Brighton in the South-East; and numerous other towns either as final destinations or as stops along the road. Over fifty of these services ran daily: the remainder, six, four or three days a week. Between Bristol and London Cary shows thirteen services, most of them daily, and some apparently employing two or three coaches. Between Bristol and Bath coaches ran all day from seven in the morning until eight in the evening.

The roads under McAdam's charge as surveyor to the Bristol Trust included all the important routes radiating from the city: the road to the Severn crossing at Aust near the site of the present bridge; the roads towards Gloucester and the Midlands; roads towards Bath and London; roads through the Somerset coalfields towards Wells; the Exeter road as far as Bridgwater. West of the lines of road towards Aust and Exeter – that is, towards the coast of the Bristol Channel, remote, marshy and flat – the Trust had no responsibilities, for in that direction there was little traffic and there were no turnpike roads.[5]

The mileage under McAdam's charge rose from 149 in 1816 to 178 after the trustees obtained a new Act in 1819.[6] Ordinary surveyors often had no more than twenty or thirty miles to look after: many much less. But McAdam was no ordinary surveyor, as the commissioners of the Bristol Trust made plain by his salary and the expenses he was expected to find from it – £400 for the first year rising to £500 from which, he said,

'I am obliged to keep a carriage and three horses and servants
. . . which I reckon cannot be done with taxes under £200.'
His area covered a radius of about thirty miles from his house.[7]

'There never was such an officer before', McAdam said of
his own appointment, which carried the title General Surveyor.
Before his time there were fifteen sub-surveyors, but no one
directly responsible for supervising them, and they were paid a
total of £672 14s (£672.70) a year. McAdam considered all
of them incompetent and most of them dishonest. He cut the
number to ten of his own choosing and paid the ten, he said,
£100 a year each, though on this point, as so often with
McAdam's statements, there is some conflict of evidence. He
thus demonstrated the value he placed on good management
by reducing the number of managers whom the trustees
employed and raising their salaries.[8]

The implications are clear and McAdam lost no opportunity
of making them explicit to any audience that would listen or
read. In the best known of his publications, *Remarks on the
present System of Road Making*, he emphasized the importance
of financial control and deplored the lack of it : 'It appears . . .
that the sum annually raised for the use of the Roads exceeds
the neat revenue of the Post Office; yet is this very large sum
expended through the hands of the lowest rank in society,' –
he meant the surveyors – 'under an appearance of control;
which equally deceives the public and deludes the expectation
of those who conduct the general business of the Roads' – that
is, the trustees.[9]

The final responsibility lay with the trustees, drawn from a
far more influential class in society than the surveyors, and
towards trustees, in the *Remarks*, McAdam found it politic to
be polite. Not so in *The Management of Trusts*, published in
1825, in which it can hardly be doubted that he revealed his
true opinions. 'As long as money can be found to expend,' he
said, 'so long Trustees continue to proceed with the old
defective system, under their own uncertain direction : and it
is not until the roads are in ruins, and the funds totally
exhausted, that they are induced to apply for a better system of

management. No sooner are the roads put into good order . . .
than they are eager to return to the former course . . . There are
already several instances of Trustees having reduced the Roads
a second time to bankruptcy and ruin, and of their making a
second application' – to McAdam, of course – 'for assistance.'
That was only a footnote. In the body of the text, attacking the
general state of the law on turnpike trusts, a favourite target,
he laid into the powers granted to trustees : 'In the regulations
for maintaining the Public Roads, and expending the vast
revenue assigned to them, the power to do evil is as un-
restrained and unlimited, as the temptations are numerous :
while all really patriotic exertions are cramped and paralyzed
by the pressure of an injudicious system of laws, and by the
host of enemies which is constantly opposed to all good management; from the indulgence afforded by a state of anarchy and
misrule, to the gratifications of vanity or of self interest.'
Drawing apparently on his own experience, which we have yet
to consider, he went on to say that in general, where an
attempt had been made to set up 'a more efficient and creditable
executive', opposition had been 'violent, powerful, and frequently successful, from the narrow and jealous feelings of
individual Trustees, and from the interested views of worthless
public servants, and those who support them.'[10] McAdam was
not universally popular.

From observing this lamentable state of affairs McAdam
drew the conclusion that since trustees were at best part-time
amateurs and at worst corrupt, they had better be provided
with professional assistance in the person of 'an executive
officer whose sole attention should be given to the business;
whose services should be amply remunerated, and of whom the
Commissioners might *of right* demand an account of the
manner in which their orders were carried into execution; who
should examine and audit the accounts of the Sub-surveyors;
compare them with the work performed, and certify them, if
approved, to the Treasurers.'[11]

If the trust were large, this official's time would be fully
taken up. His duties would be part technical, part financial, and

McAdam was precise about how they should be carried out. 'He must direct the execution of the repairs, and alterations of the road, when ordered by the Commissioners; and he must controul the contracts, and other agreements entered into by the Sub-surveyors, so as to prevent unnecessary expence; he must examine all work performed, to see that it is corresponding with contracts, and generally keep a vigilant superintendance over the persons employed under him. Accounts of all expences incurred should every second week be delivered by the Sub-surveyors into his office in duplicate; after examination, one copy to remain in the office, the other certified, to be sent to the Treasurer, upon which payment may follow.'[12]

Finally, such officers must be gentlemen:

'Much must depend on the selection of the officer to whom this charge is committed; he must have a considerable share of general information respecting country business; the subject of road-making ought to have been well considered by him; his station in society should be such, as to secure to him the support and confidence of the Commissioners, while it commands the obedience and deference of the subordinate officers.'

Such a treasure, if he could be found, would be expensive, but the expense would be fully justified:

'Skill and executive labour must be adequately paid for, if expected to be constantly and usefully exerted; and if so exerted, the price is no consideration when compared with the advantage to the public.'[13]

McAdam's proposals for surveyors of this class – general surveyors to whom sub-surveyors would be answerable for the detailed care of the roads – arose from his horror at the spectacle of turnpike debts of more than £7 million being left to the uncontrolled mismanagement of unpaid amateurs with no properly trained full-time officials to assist them. 'England alone', he pointed out, 'is divided into 955 little Trusts which may be considered, in fact, as hostile to each other; while it is evident that unity of action is of vital importance among Commissioners of the same branch of the public service . . . it will hardly be deemed inexpedient to recommend some central control over the District Commissioners, that may have the effect of regulating the eccentricity of their measures, as well as giving their views, in many instances, a better direction. This central control will be most beneficially established in each county, under such regulations, and with such powers as the wisdom of Parliament may deem most effectual.'[14] To make this plan work, there should be a general surveyor in each county, presumably with powers of inspection over each separate trust. No such system was ever set up by law, though McAdam and his son James tirelessly advocated something of the sort, but we shall see that McAdam and his sons came close to establishing it unofficially, in some parts of the country, by acting as general surveyors for groups of trusts covering wide areas and long continuous stretches of some of the 'great roads' of the kingdom.

In his specifications for a general surveyor, McAdam was drawing a self-portrait. More than that, he was drawing the portrait of a figure fast rising to importance in the expanding economy of 19th century England: the professional man. A road surveyor qualified for the position outlined by McAdam would take his place alongside the attorney (who was beginning to prefer to be called a solicitor), the apothecary (who, to the physicians' fury, was more and more confidently aspiring to perform some of their functions), the surgeon, the architect, the civil engineer : all those pushing middle-class men who were seeking to establish themselves in public esteem and secure a comfortable living by a combination of superior technical knowledge, improved ethical behaviour, and enhanced social standing. At one remove again, he was describing a modern manager, competent to execute policy, to watch costs, to direct the work of subordinates, and accustomed to take his reward by salary, not by the profits of ownership.[15]

McAdam was equally specific and demanding in the requirements he laid down for 'procuring a more skilful set of sub-surveyors' in the lower branch of the profession. It was a business, he said, which could not be taught from books, but could 'only be acquired by a laborious practice of several months and actual work upon the roads, under skilful road-makers. Young men who have been accustomed to agricultural labour are fittest to be made road-surveyors, as their occupations have given them opportunities of being acquainted with the value of labour both of men and horses.' Elsewhere he described them as being 'selected from the most respectable yeomanry'.[16] Thomas Brassey (1805–1870), the almost legendary Victorian railway contractor, seems to have been just such a young man as McAdam sought. He was the son of a well-to-do yeoman farmer and after Chester Grammar School he was apprenticed to a land agent and surveyor. 'One of his first qualified jobs was to work as assistant to Telford, on the early Cheshire turnpike roads.'[17]

McAdam's appointment at Bristol came after the Commissioners, as the climax to a period of growing uneasiness about the state of their finances and of their roads, found themselves under indictment at the instance of the Postmaster General.[18] McAdam, ten years later, described the situation with relish: 'A large increasing debt, obscure and perplexed accounts, dilapidated funds, peculation, fraud, and ignorance among the inferior officers, ineffectual control, and misdirected authority among the higher; and the Roads, for which such large sums were drawn from the Public . . . were nearly in ruins, and some of them under notice of indictment.' In all McAdam's published work, allowance has to be made for self-advertisement and for a somewhat carefree attitude towards facts, but probably the picture he drew was not far wrong, for the Postmaster General was normally peaceably disposed towards turnpike trusts, unwilling to bring proceedings unless, as the Superintendent of Mail Coaches put it in 1819, the state of the roads constituted 'a very considerable evil'.[19] The Bristol Commissioners responded as turnpike trustees commonly did when a problem arose. They appointed a committee.

The Committee found that it was costing a great deal of money to keep the roads in their deplorable state: more, at any rate, than the tolls were bringing in, for between 1802 and 1812 the trust's debt had risen by 50 per cent, from £22,827 to £34,565. It was later found to have doubled, for according to a Memorandum to the Committee's report an extra sum of £10,400 was 'afterwards stated' from the Winford Division, bringing the total debt in 1812 to £44,065. Accountancy, as McAdam had realized, was not in these years the Bristol Trust's strong point. The Committee could, however, recognize insolvency when they saw it and they recommended erecting additional tollgates, 'every toll gate to receive the full tolls payable at present' (McAdam in 1823 said they recommended doubling the tolls), and they added 'that the measure they recommend must be adopted, or the roads which are now insolvent must go to utter ruin.' They recommended also, 'with extreme pain and reluctance', that 'an immediate stop be put to all farther extensive improvements'. They declared it their opinion 'that the business relating to the roads has not been well managed; but appears always to have been in a state of obscurity and perplexity, without method or regularity: That the system of management has been radically bad, without head or focus to give it effect.'[20]

This was the message McAdam had been waiting for. He had already reported on the state of the Bristol roads in the previous June and July, and on 4th December 1815 twenty trustees, of whom he was one, met at Bristol Guildhall to hear his report read. They went on to recommend him for the newly created post of General Surveyor, with authority over all the surveyors in the districts into which the roads of the trust were divided, his appointment being for three years at the salary of £400 a year already noticed (see p. 43). It may not be without significance that the proposer of the motion had a Scottish name: Thomas Graeme.[21]

The General Surveyor's duties were set out at length. He was to attend all General Meetings of trustees and report on the state of the roads 'with his observations thereon, and opinion as to all necessary reparations or improvements'. He was to 'form the specification of all contracts . . . to make himself fully acquainted with the state of repair and the local circumstances, in respect to materials, carriage and labour of each separate Road : – to visit each Road . . . as frequently as time will permit, and direct and superintend all the particular Surveyors appointed by the separate Meetings (in the various districts), and on the first Monday in March, in every year, to present a written Report to the General Meeting of the state of each separate Road under his Survey.'

On these terms McAdam at last got his chance to put his theories into practice. According to his own account, published in 1825, he set about his duties against strong opposition. 'So great', he said, 'was the alarm spread among the lower orders, the farmers, the tradespeople at the commencement of the year 1816, that it required no common degree of firmness to persevere in a plan which had no precedent, nor even the experience of a single season to support it. It was, therefore of vital importance to the success of the measure, to show from the first its decided superiority : and the most zealous exertions and utmost resources of the original proposer were employed to second the laudable designs of the Commissioners.'[22]

The alarm McAdam's appointment caused in certain circles is understandable, for he descended on the comfortable world of the surveyors and their cronies like an angel of wrath. He made 'a most careful and minute examination . . . of the whole roads of the district', taking account of 'their strength . . . as respected quantity of metal; the quality and size of stone used; the manner of application and consequent condition of each road for public use; methods of draining; manual labour; carting; prices paid under each head, and every other particular.' He found, as no doubt he expected, 'the ignorance and incapacity of Surveyors; the want of all effectual control over the most lavish expenditure; the inexplicable confusion of accounts; the absence of all system or scientific mode of constructing the roads, every part being differently formed, and managed according to the varying opinions of individuals; and that these causes were amply sufficient to occasion the alarming results complained of in the Report of 1815.'

In February 1816 he began to lay the foundations of a system of management based on accurate factual knowledge. He wrote to each surveyor demanding a fortnightly report showing the labour employed and the wages paid; contract work and its expense; work done by each labourer individually; employment of 'Waggons and Carts' and the expense; 'the quantity of stone carried to the road, or laid on it in the two weeks, distinguishing Statute Labour'. He followed that with a printed questionnaire asking each surveyor how long he had been employed and at what wages; the length and breadth of his division and the depth of stoning on it; whether he was under contract; how much stone there was in quarries, by the side of the road, quarried by contract; what cartage was done by contract; how much stone was used every year; how much stone was broken by contract; what size of stone was used in the last repair; how many men were employed. 'These papers', he noted ominously, '. . . were deposited in the Office, as a check upon the irregular conduct of the then Surveyors, some of whom were strongly suspected of practices very inconsistent with their duty.'[23]

The surveyors 'with a few honourable exceptions' did all they could to hinder McAdam. He responded by bringing in his relations: the first example of a practice by which, in the next few years, his system was rapidly extended. He does not name those he called in, but other evidence suggests that they were his son John Loudon Jr, who became Joint General Surveyor with him in 1824, and his grandson John Loudon James Saunders, born in 1803, the son of McAdam's eldest daughter Ann (1779–1841) and Captain James Saunders RN, whom she married in 1801.[24]

The McAdams, according to J L McAdam, themselves took over the duties of surveyors 'who in many instances were detected instigating the labourers to injure or retard the work'. At length they succeeded in 'procuring a body of respectable and efficient Surveyors', but 'as the system . . . was entirely new and unknown, it was necessary for us to instruct them in the very rudiments of their calling'. The training of surveyors, like the employment of relations, was to be repeated many times as macadam spread across the country. McAdam, writing of these early Bristol recruits, praised their 'zeal and fidelity, after being carefully trained and instructed by myself and my family'.[25]

McAdam persuaded the Commissioners to disregard the recommendations of the Committee of 1815. No additional gates were put in, and as soon as he could after necessary repairs had been done, McAdam went in for widespread improvement. As a measure of first aid, in his first year, he had the surface of about 10 miles of road lifted and relaid, the stones having been broken small, and as time went on the same process was repeated over all the roads of the trust. He was careful to point out that this was purely emergency surgery, not the whole of his system, as some of his critics assumed, and he went on to say that before a road could be considered properly built all the stone in it should be broken small and laid directly on the soil – well drained – without a paved foundation. On this point, as we shall later see (see p. 113), he fell foul of Telford, who insisted on a hard road-bed of large stones.[26]

In June 1817, about eighteen months after his appointment, McAdam reported that no part of the Bristol roads was in a bad state and numerous improvements had been made, and all this for an outlay which was steadily falling, year by year. In 1815, 'previous to the alteration of management', £15,685 2s 1d was spent on the Bristol roads; in 1816, £14,387 5s 1d; in 1817, £13,470 4s 11d. In 1818 so much had been saved that the commissioners were able to spend a great part of it on improvements instead of putting it all to the reduction of debt.[27] Nevertheless the debt went steadily down.

In 1823, for the benefit of a parliamentary Select Committee, McAdam summed up his achievements at Bristol:

'The debt of the road at the settlement in March 1821 . . . was £40,293; the report of the Committee in 1815, shows the dreadful state in which the roads were, with a debt of £45,236; and shows that the Committee could suggest no means of extrication, except double tolls, and abstaining from permanent improvements. Instead of that, I took possession of the road at the old tolls; put them all in order; proceeded on with all the permanent improvements. I have already laid out, on an average, £3,000 a year in permanent improvements, and I have decreased the debt . . . I lowered a great number of hills, lengthened about twenty bridges, widened the whole of that part of the road between Bath and Bristol that is in our district: I widened several other lines of the road. I have taken down the famous Black Horse hill, which has been a stumbling block for ages. We have made twelve or fourteen miles of new roads entirely . . . out of this money. It is partly made new, and partly parish roads, widened and made turnpike.'[28]

He claimed also that expenditure for ordinary repairs had dropped since 1821 from about £15,000 to £12,000 a year: 20 per cent.

The essence of McAdam's estimate of his work at Bristol, therefore, was that he had delivered to the Commissioners the double benefit which he always insisted was the unique merit of his system: better roads for less money.

How had he done it? The members of the Select Committee of 1823 thought he owed a good deal to the collapse of agricultural wages after 1815. He had entered on the Bristol job, they pointed out, 'at a very favourable time, on account of the reduction in the price of labour', suggesting that wages since the war might have fallen by as much as 30 per cent. McAdam disputed that figure, but he agreed that he was paying day labourers twelve shillings a week, whereas they would have had fifteen shillings during the war. Since, on his own showing, labour on the roads in 1821 accounted for 31 per cent of the total expenditure for repairs, it may be conceded that the Select Committee were nearer the mark than McAdam liked to admit.[29]

At the height of their activity McAdam and his sons, like other employers, took full advantage of the miserably low rates of agricultural wages prevailing during the twenty years or so following the Napoleonic wars. Moreover, in keeping with the traditional view that the building and repair of roads called for no special skill, road work was generally considered fit only for the lowest class of labour : those too poor to buy themselves out of compulsory 'statute labour'; aged, infirm; paupers; unemployed labourers.

No surveyor would employ statute labour if he could avoid it, preferring always commutation money, even at a very low rate. 'If it were only half of the real value of the work,' McAdam told the 1819 Select Committee, 'I should think the roads would be more benefited by it in general throughout the country.' Paupers, whether hired on contract from the overseers of the poor or recruited directly, were nearly as bad. James McAdam, in 1824, complained of 'want of skill and knowledge in the overseer', and his father, in 1819, thought the price of pauper labour too high. 'Is the pay of those men proportionably low with their abilities to work?' he was asked, and he replied : 'I have found that those poor, miserable men, who can do very little, have been getting considerable wages, and in that way a considerable sum has been wasted.' Parishes, of course, were always anxious to get the poor off the rates, but McAdam, not unreasonably, objected to the road revenue being 'made to act as a poor fund'.[30] He discovered, as others have discovered since, that job creation is not sound economics.

Able-bodied farm labourers, through lack of employment, might be reduced to working on the roads as an alternative, or sometimes in addition, to 'going on the parish', and the McAdams naturally took credit for relieving distress. 'A vast number', said James McAdam in 1824, would be out of work between October and April, when as a consequence, most of the road work would be done. In the summer, especially at harvest time, labour would be scarce. Whole families would be employed, and James quoted 'a great number of instances of men with no less than ten children, and the wife, being wholly employed upon the turnpike roads'.[31]

Wages varied widely, being generally linked to the local rate for farm labourers. In the home counties, Lincolnshire and

Hampshire, James, in 1824, was accustomed to paying between 9s (45p) and 14s (70p) per week: generally 10s to 11s or rather more in London. A man with a wife and eight children, he thought, might earn 30s to 35s (£1.50 to £1.75). In 1836, a foreman working for the Hauxton and Dunsbridge Trust near Cambridge had 12s plus an allowance for harvest money. By 1847, labourers' wages on that trust's roads were being fixed at 12s. Generally speaking, wages were highest where there was competition for labour from industrial towns, as in the North: lowest in the purely rural counties of the South, South-East and South-West.[32]

Neither James nor his father would pay by time rates if they could help it. Both much preferred payment for accurately measured piece-work. They got more out of the men that way, and induced them to compete with each other. The men, for equal and opposite reasons, preferred day-work and would even refuse piece-work if they thought the parish would relieve their families. J L McAdam, nevertheless, managed to get all his men at Bristol on to piece-work. Asked his opinion of 'the common mode of employing paupers by day-work', he replied:

'It may have the effect of relieving the parishes, but I should think it a very bad mode of mending the roads; inasmuch as these men, when they have got day-wages, will do very little, and for that reason I employ all our men on piece-work; we have two hundred and eighty labourers in the district of Bristol, and they are almost all on piece-work; it is very seldom we employ men by the day.'[33]

McAdam's brisk, hard, unsentimental efficiency delighted the Bristol commissioners. His appointment was at first for three years, so that it was due to run out at the end of 1818. At a meeting on 7th December of that year the commissioners decided to renew it for three years, at the same salary, and resolved unanimously:

'That the thanks of this meeting be given to Mr McAdam for the zeal and ability with which he has executed the very

arduous duties of his office, from which it appears to this
meeting that the most important advantages have resulted to
the roads under his care.'[34]

By this time McAdam, at the age of 63, was a national
figure, for his success had attracted attention far beyond
Bristol. Turnpike trustees throughout the country, struggling
with badly kept roads, apparently unmanageable debts, incompetent – sometimes dishonest – surveyors, were sending to
McAdam for advice, reports, practical help in training surveyors, sub-surveyors and workmen.

McAdam was very willing to oblige, having an evangelical
zeal for the advancement of his system. He never lacked support
at high political and social levels, and he was launched upon his
missionary endeavours after coming to the notice of a grandee
with access to the inmost circles of power and patronage:
Thomas Pelham, Earl of Chichester (1756–1826). Chichester,
a politician rarely out of office from 1782 until his death, became
Postmaster-General in 1807 and held the appointment for the
rest of his life, for some years jointly with the first Marquess of
Salisbury. In his official capacity, he was the greatest user of
turnpike roads in the country.[35]

Chichester told the Select Committee of 1823 that he first
made McAdam's acquaintance about 1816 or 1817, when there
was 'a great desire in the public that the mails should be
accelerated, and a great complaint of the general state of the
roads, particularly by the contractors for mail coaches'.[36]
When, in response to this demand, the Post Office authorities
wanted to provide a faster service to Bristol they were told
'that if the roads about London were as good as they were about
Bristol . . . the contractors could very easily accelerate their
pace'.

Francis Freeling (1764–1836), a Bristolian and a Post Office
official celebrated both as an administrator and as a man of
taste, being a Fellow of the Society of Antiquaries who formed
'a curious and valuable library',[37] thereupon wrote to the
postmaster of Bristol, who praised McAdam's work for the

Bristol Trust. At about the same time Sir John Coxe Hippisley (1748–1825), who had heard of Chichester's enquiries, wrote to tell him of improvements made by McAdam to the road passing Hippisley's house at Ston Easton in Somerset. Hippisley had retired there, having acquired the house through his second marriage, after diplomacy in Italy, 'offices of trust and importance'[38] in India during the war with Haidar Ali and Tipu, and successful negotiations for the marriage of the Princess Royal to the Duke of Württemberg. He devoted himself to the orthodox business of a country gentleman, including service as a trustee of the Bath roads.

McAdam, thus recommended, came to see Chichester, who must have been an enthusiast for turnpike road reform in his private capacity also, for he had become surveyor to a trust near his house in Sussex, an extraordinary step for a nobleman and a great landowner. He and McAdam, exact contemporaries, apparently got on well together. After 'a good deal of discussion' Chichester had the local flint road lifted and relaid according to McAdam's plan, 'which afforded some amusement to the country people, who rather laughed at the thing, and thought that those little hammers and those little pebbles were a great waste of money, and that the digging up of the old road was also very absurd'. The men found the work difficult to learn, so McAdam sent for one of his foremen from Bristol 'and we then employed any men that offered'.

'In the course of conversation with Mr McAdam', Chichester went on, 'it occurred partly to him, and partly to myself, that the great thing to be wished for was to have the management of the roads put into the hands of a man of science, and a gentleman, in order to correct the sort of neglect and jobbing . . . too frequent in works of that kind. Farmers could not possibly pay much attention to it, and if another man was appointed at a salary, it generally was a matter of favour.'

This was perhaps the origin of the idea of a 'general surveyor' which McAdam later elaborated on in his *Remarks*, for Chichester proposed uniting the roads of five trusts round Lewes under a general surveyor to be found and instructed by McAdam.

There was no authority in the relevant Acts for this consolidation, but the thing was done by general consent of the trustees, and a man called Campbell – no doubt a Scot – who had worked under McAdam at Bristol was put in as surveyor-general, with a salary of £150 a year. After a winter of his exertions, the joint trustees were so well pleased that they doubled his pay.[39]

This seems to have been the start of what swiftly became a widespread consultancy practice. By the time McAdam gave evidence before the 1819 Select Committee he 'had been sent for and consulted by thirty-four different sets of commissioners, and as many different trusts, and in thirteen counties, to the extent of 637 miles, all of whom have been making improvements, and I have had many sub-surveyors instructed and sent to various parts of the country, at the request of commissioners; many surveyors also in the neighbourhood where improvements are making, have availed themselves of the opportunity of having instruction. Thus the surveyors of Southampton and that neighbourhood have attended to what is doing at Salisbury and Wilton; thus the surveyors at Kingston and Guildford have profited by the improvements at Epsom in Surrey.' By 1823 he claimed to have visited and advised seventy trusts in twenty-eight counties.[40]

To the surprise of his contemporaries, perhaps to the incredulity of some, McAdam maintained that he gave his services as a consultant free of charge, sometimes without even a claim for expenses. 'In every case', he explained, 'where I was written to for advice and assistance, . . . it was preceded by a letter to know what the expense would be, and upon coming to every trust afterwards, I found that there was a party adverse to my being consulted, and adverse to . . . the application of my system; that they had always opposed my being sent for, until a letter came from me to say that it would be no expense at all . . . and I believe that if I had not sent such answers . . . I should not have been consulted by seven of those trusts; and my belief is, that if I had made it a money-making speculation, I should have strangled the business in its birth.'[41]

A cynic would no doubt say that free consultancy was bait laid out to draw lucrative business towards him or his family later. Sir Alexander Gibb (1872–1958), writing in 1935 from a position of immense eminence in the engineering profession, considered McAdam's behaviour most unethical – 'he humbly disclaimed all right to the title and honour of engineer, and accordingly felt himself unrestricted by either the etiquette or the ethics of the profession. He associated with himself a number of his sons and other relations, and together they scoured the country, disregarding all existing connection of engineers with the various road authorities.'[42] Chichester, who had the advantage of personal acquaintance with McAdam, took an altogether more charitable view:

'I should wish to say, when I am asked about remuneration, that Mr McAdam coming to me in the manner that he did, as a gentleman, and as a brother surveyor, it did not occur to me to offer anything as compensation; at the same time, as the thing went on, it certainly did appear to me that it would be necessary that he should be paid somehow or other, but he always declined it, and said that his main object was to introduce the system, and he did not know very well how to make any charge, and I confess that I was of the same opinion; it seemed to be the sort of benefit that you could not reduce to figures, and I do not know how the trustees of our five roads could have given Mr McAdam a gratuity out of those funds; it would have been difficult.'[43]

In August and September 1817 McAdam laid 200 yards of road near Blackfriars Bridge in London and 180 yards near Westminster Bridge using, experimentally, gravel dredged by steam engines from the river-bed and pebbles from the coast, instead of the more usual 'gravel of the country'.[44] In several years he had a great deal of parliamentary business, partly giving lengthy evidence before select committees, partly promoting the new Act for the Bristol Trust which passed in the session 1819–1820. His advice was repeatedly sought by the Post Office, and on one occasion he was asked to go to

Lancashire to investigate postmasters' complaints that their horses were being destroyed by paved roads. In 1817 he was nearly appointed surveyor to the Bath Trust, proposing to hold the appointment along with his post at Bristol, but he was narrowly defeated after a spirited campaign in local politics (see p. 130). With all this and his consultancy service, he must have been travelling almost as much as in the years when he was inspecting the roads, and it all had to be fitted round his duties as the salaried surveyor to the Bristol Trust – duties to which, as the terms of his appointment (see p. 49) clearly show, he was expected to devote the whole or nearly the whole of his time.[45]

Faced with this situation, he called up family reinforcements. 'I have called from Scotland my eldest son, now 34 years of age', he told the Bath trustees in March 1817, characteristically getting his son's age wrong – William was about 36. William's own account of the matter was that he arrived in England in 1818, so perhaps he needed persuasion to give up his job with the British Tar Company or had to give lengthy notice before he did so. However that may be, the second son, James, took up a surveyorship at Epsom, originally offered to his father, in December 1817.[46] McAdam's sons were accompanied or followed, with varying degrees of success, by a nephew, four grandsons and at least one connection by marriage. The impact of the McAdam kindred on the turnpike roads will be described in succeeding chapters.

In the summer of 1819 McAdam was invited by Robert Dundas, 2nd Viscount Melville (1771–1851) to go to Scotland for consultation about the roads of the Edinburgh County Trust – 273 miles organized, in the Scottish fashion, in ten semi-autonomous districts. Melville was a great man and McAdam perhaps felt flattered. He got leave from the Bristol trustees, went to Edinburgh, inspected the roads, and reported in his usual terms, criticizing the way they were constructed, advising a good deal of lifting and re-laying, attacking the local surveyors and suggesting the appointment of a general surveyor. His report seems to have been well enough received centrally, but the ten districts, who would have to pay for it all, were not consulted.[47]

The trustees asked McAdam – who else? – to find a general surveyor and he nominated his 31-year-old nephew Lieut-Col James Shaw (1788–1865), who after distinguished service in the Peninsula and in the Waterloo campaign was on half-pay. He was contemplating marriage, he needed an income, no doubt his mother was pressing. What more natural than that his uncle should put him forward? He knew nothing of road surveying, but he was given three months with his cousin James and with Lord Chichester to put that right. As a dashing Regency soldier, of course, he had to have more money than the common run of surveyors, so a salary of £700 a year was promised: far more than his uncle or any of his other relations ever had from a single trust.

The appointment was a blunder. The age accepted nepotism readily enough as the natural way of doing things, but this was a particularly blatant example and for the Edinburgh trustees, especially on the outskirts, it was compounded by the lack of consultation, by the exorbitant salary, and by Shaw, an outsider, having been thrust on them by another outsider. Moreover he arrived in Edinburgh accompanied by a cousin, George Pearson, and by two English sub-surveyors who were to have a hundred guineas – *a hundred guineas* – a year each.

McAdam's methods were fairly readily accepted on the Edinburgh roads, but his nominees were not. After two years Col Shaw came up for re-appointment. He was dismissed. He was re-employed in the army, added his wife's name to his own, and after a long and honourable career died as General Sir James Shaw-Kennedy KCB. In the Dictionary of National Biography there is no mention of the Edinburgh episode.

J L McAdam was not so busy with turnpike roads as to neglect the possibilities of railways. On 27th December 1824 a meeting of merchants at the London Tavern in Bristol resolved to form the London & Bristol Rail-Road Company with a capital of £1.5m. Among the Directors, several with Scottish names, was J L McAdam. He was also appointed engineer to lay out the line: an appointment which shows how informal professional qualifications were in the early 19th century. In a little more than a fortnight he produced a plan for a railway from Bristol to Brentford, largely following the route later adopted by I K Brunel along the Vale of the White Horse and then with two alternative routes along the valley of the Thames. The railway seems only to have been intended for goods traffic, and McAdam proposed also a turnpike road 'passing through the towns instead of near them, as the Railroad must necessarily do'. The Cotswold ridge he described as 'a small swell of the country near Dodington'. 'We can only assume', says the historian of the Great Western Railway acidly, 'that he took no levels', and indeed the whole plan, produced so quickly, sounds so perfunctory that it seems doubtful whether it was ever taken very seriously. It was part of the earliest boom in railway shares which collapsed in the general crisis of 1825, and after the early part of that year no more was heard of it.[48]

By the autumn of 1824 a considerable body of Bristol trustees was dissatisfied with McAdam's frequent absence from his post. Mrs Pember says that in September 1824 a motion to dismiss him failed by only one vote, and in the following month a committee of trustees took evidence from eight of the sub-surveyors and six of the district treasurers (distinguished as Esq. against the surveyors' Mr), clearly with the intention of making a case against him. McAdam was not invited to attend or give written evidence.

The surveyors were in a delicate position. If they disliked McAdam, here was their chance. On the other hand, as they well knew, he could put them in the way of bettering their position. It says a good deal for his standing with them that their evidence gives the impression that one and all were uneasy and unwilling to incriminate him, though some were more forward in defending him than others. The treasurers, on the other hand, were forthright. Two were satisfied with McAdam: the other four, in one form of words or another, asserted that he had not attended to his duties as he should have done.[49]

The Committee, as McAdam had no doubt foreseen, found against him, reporting 'their opinion that the Absences of the General Surveyor have been such as to be entirely inconsistent with the discharge of his duty to the Trust, and decidedly contrary to the terms of his original engagement'. They sent him a copy of the evidence and informed him 'that he was at liberty to adduce such evidence, and offer such explanations, as he might think proper; and that the clerks were directed to summon any witnesses whose attendance he might require'.

McAdam never shirked a fight, but he had a strong regard for his own dignity. He replied on 14th November 1824 from Penrith – as follows:

'I must decline making any observations, or any other interference with proceedings taken in secret, or with evidence which I was precluded from hearing delivered.'[50]

He followed that up with a spirited *Narrative of Affairs of The Bristol District of Roads from 1816 to 1824*, preserved in Bristol Record Office, which was printed for him by J M Gutch, probably a trustee or relation of a trustee, since the name Gutch appears in a list of trustees attending a general meeting. At a meeting of seventy-one trustees, with the Mayor of Bristol in the Chair, on 2nd December 1824, the chairman of the investigating committee moved for McAdam's dismissal, but by forty-five votes to twenty-six his motion was amended to a vote of censure in fairly mild terms.

Controversy followed in the local press, provoking McAdam at one point to threaten a writ for slander, and then in March a committee of trustees was appointed to examine an offer from a man called Lucas to repair the roads by contract: an offer which no doubt McAdam regarded as an affront. In spite of a report in his favour by the committee, in May, he sent a letter of resignation in June, to take effect on 25th March 1826, and it was accepted in July.[51]

Resignation may be an admission of weakness or a show of strength. In McAdam's case it seems to have been the latter. His friends moved in September 1825 to re-employ him, but the motion was put off until the December meeting. Then, on 2nd December, with the Mayor once more in the Chair, the trustees invited him to take office again from the date on which he was proposing to lay it down. They asked him to give one fortnight a quarter to the Trust's business, in return for £250 a year.

McAdam's triumph went further. In 1827 the trustees approached him 'calling for his assistance after having abandoned his System'. In reply he favoured them with a long letter running through his usual points about the incompetence of most surveyors, the necessity of appointing a General Surveyor *and paying him adequately*, and his own activities at Bristol. 'Under this system,' he said, 'the roads of the Bristol Trust became a pattern that has been imitated in some measure by every part of the Kingdom.' He went on to say that it did not become him 'to account for the motives that induced a Number of the Trustees to disturb this system, and finally to break it up', and to claim that 'consideration for the public interest weighs more with me than private Emolument'.[52] It sounds as if an attempt had been made, once again, to put the Bristol repairs out to contract, and had failed. McAdam's letter is typical of his style and he must thoroughly have enjoyed writing it. A copy found its way into the archives of the Exeter Trust, to which his son William and his grandson Christopher were surveyors, and there it has been preserved.

John Loudon McAdam remained General Surveyor of the Bristol Roads, with his son Loudon as his colleague, until he died, and as if to emphasize his independence he took on seventeen other surveyorships as well (see p. 74). After his death, Loudon succeeded him and in his turn held office for the rest of his life. At Bristol the McAdams' victory could not have been more complete.

Surveyor of the Bristol Roads 1816–1825

REFERENCES

The principal sources cited are:

John Loudon McAdam: *Remarks on the Present System of Road Making*, 7th Edn 1823, cited as *Remarks*; *Narrative of Affairs of the Bristol District of Roads from 1816 to 1824*, 1825, cited as *Narrative*; *Observations on the Management of Trusts for the Care of Turnpike Roads*, 1825, cited as *Management*.

Parliamentary Papers: Report from the Select Committee on Mr McAdam's Petition 1823, cited as SC 1823.

Robert H Spiro Jr: *John Loudon McAdam – Colossus of Roads*. Unpublished PhD thesis, Edinburgh 1950, cited as Spiro.

1. SC 1823 11.
2. Devereux, *John Loudon McAdam* 129.
3. Spiro 143; see also 156 for McAdam's civic activities.
4. Angus Buchanan and Neil Cossons, *Industrial Archaeology of the Bristol Region*, David & Charles 1969, 16 and generally.
5. (4) 177–8. Brenda Buchanan, 'Turnpike Roads in a Regional Economy'.
6. SC 1819, evidence of J L McAdam 4iii19. *Management* 29.
7. (1) 13, 16, 18.
8. (1) 17, but see also evidence of J Light (£2 16s. a fortnight), Edward Whitting (£80), James Stokes (£150) in *Narrative* 36, 33, 31.
9. *Remarks* iv 17–18.
10. *Management* 21n 31–2.
11. (9) 18–19.
12. (9) 19.
13. (9) 20–1.
14. (9) vi.
15. Cf Alfred D Chandler Jr, *The Visible Hand*, Harvard UP 1977, Part One.
16. (6); (9) vi.
17. R K Middlemas, *The Master Builders*, Hutchinson 1963, 33.
18. (1) 14.
19. (10) 10; (6), evidence of Charles Johnston 2iii19.
20. *Narrative* 4; see also (1) 14 for slightly different figures; (10) 13–17.
21. Bristol Trust Minutes 4xii15, quoted (10) 119–21.
22. (10) 12–13.
23. (10) 17–19, 14, Appendix 5.
24. (10) 15–16, 129 (Bristol Trust Minutes 28iv24); *Narrative* 29 (evidence of George Down); family details from Spiro.

25. (10) 15–16.
26. (10) 21–5.
27. (6), evidence of J L McAdam 9iii19.
28. (1) 16.
29. (1) 18.
30. As 27; Report from the Select Committee on Labourers' Wages 1824, 14.
31. SC of 1824 as (30) 11.
32. (31) 5 11–14; Hauxton & Dunsbridge Trust Minutes 22i36, Cambridge University Library MS 6015, also 22i47.
33. (6) 9iii19.
34. (6) 9iii19.
35. DNB.
36. (1) 18.
37. DNB.
38. DNB.
39. (1) 18–19.
40. (6) 4iii19; (1) Appendix Q.
41. (1) 50.
42. Sir Alexander Gibb, *The Story of Telford*, Alexander Maclehose 1935, 178–9.
43. (1) 20.
44. (6) 9iii19.
45. (1) 55.
46. (1) 97; 41, 49.
47. Narrative of events in Edinburgh condensed from Spiro 328–53.
48. (2) 111; E T MacDermott, *History of the Great Western Railway*, 2 vols privately printed by the GWR 1927, I 1–2; Harold Pollins, *Britain's Railways*, David & Charles 1971, 22.
49. (2) 114; *Narrative* 29–39.
50. *Narrative* 40* 41* (the asterisks are part of the page numbers: the pagination in McAdam's publications is always eccentric).
51. Spiro 263.
52. Letter from J L McAdam to Bristol Trustees 3vi27, appended to the Report of a Committee of Trustees in Exeter Trust Minutes ETT 2/5, 93, Devon RO.

Ode to Mr McAdam

from "Odes and Addresses to Great People"   1825
by Thomas Hood

1

'Let us take to the road'. – *(The Beggar's Opera)*

M'ADAM, hail!
Hail Roadian! hail, Colossus! who dost stand
Striding ten thousand turnpikes on the land!
　　O universal Leveller! all hail'.
To thee, a good, yet stony-hearted man,
　　The kindest one, and yet the flintiest going, –
To thee, – how much for thy commodious plan,
　　Lanark Reformer of the Ruts, is Owing!
　　　The Bristol Mail,
Gliding o'er ways hitherto deem'd invincible,
　　When carrying patriots now shall never fail
Those of the most 'unshaken public principle'.
　　　Hail to thee, Scot of Scots!
　　Thou northern light, amid those heavy men!
Foe to Stonehenge, yet friend to all beside,
Thou scatter'st flints and favours far and wide,
　　From palaces to cots; –
　　Dispenser of coagulated good!
　　Distributor of granite and of food!
Long may thy fame its even path march on,
　　E'en when thy sons are dead!
Best benefactor! though thou giv'st a stone
　　To those who ask for bread!

2

Thy first great trial in this mighty town
Was, if I rightly recollect, upon

That gentle hill which goeth
Down from 'the County' to the Palace gate,
    And, like a river, thanks to thee, now floweth
Past the Old Horticultural Society, –
The chemist Cobb's, the house of Howell and James,
Where ladies play high shawl and satin games –
    A little HELL of lace!
And past the Athenaeum, made of late,
    Severs a sweet variety
Of milliners and booksellers who grace
    Waterloo Place,
Making division, the Muse fears and guesses,
'Twixt Mr. Rivington's and Mr. Hessey's.
Thou stoodst thy trial, Mac! and shav'd the road
From Barber Beaumont's to the King's abode
So well, that paviours threw their rammers by,
Let down their tuck'd shirt-sleeves, and with a sigh
Prepar'd themselves, poor souls, to chip or die!

3

Next, from the palace to the prison, thou
    Didst go, the highway's watchman, to thy beat, –
    Preventing though the RATTLING in the street,
        Yet kicking up a row
Upon the stones – ah! truly watchman-like,
Encouraging thy victims all to strike,
    To further thine own purpose, Adam, daily; –
Thou hast smooth'd, alas, the path to the Old Bailey!
    And to the stony bowers
Of Newgate, to encourage the approach,
    By caravan or coach, –
Hast strew'd the way with flints as soft as flowers.

4

Who shall dispute thy name!
Insculpt in stone in every street,

Two views of Clifton Turnpike on one of the Bristol Trust roads,
(Chapter III), *above* before reconstruction, looking west over the Avon
Gorge towards S. Wales; *below* after reconstruction, 1823, looking east.

City of Bristol Museum and Art Gallery

Not the way to do it, 1812. This man is breaking the stone in a standing position, with a long-handled hammer, into pieces which are much too large (pp. 37–8). *Fotomas Index*

Highgate Tunnel, more commonly called the Archway, 1822. The stone-breaker is using a longer hammer than McAdam recommended (p. 38). *Museum of London*

Ode to Mr McAdam

We soon shall greet
Thy trodden down, yet all unconquer'd fame!
Where'er we take, even at this time, our way,
Nought see we, but mankind in open air,
Hammering thy fame, as Chantrey would not dare; –
  And with a patient care,
Chipping thy immortality all day!
Demosthenes, of old – that rare old man –
Prophetically, FOLLOW'D, Mac! thy plan : –
  For he, we know,
  (History says so)
Put PEBBLES in his mouth when he would speak
  The SMOOTHEST Greek!

5

It is 'impossible and cannot be',
  But that thy genius hath,
  Beside the turnpike, many another path
  Trod, to arrive at popularity.
O'er Pegasus, perchance, thou hast thrown a thigh,
Nor ridden a roadster only; mighty Mac!
And faith I'd swear, when on that winged hack.
Thou hast observ'd the highways in the sky!
Is the path up Parnassus rough and steep,
  And 'hard to climb' as Dr. B. would say?
Dost think it best for Sons of Song to keep
  The noiseless TENOR of their way? (see Gray).
What line of road SHOULD poets take to bring
  Themselves into those waters, lov'd the first! –
Those waters that can wet a man to sing!
  Which, like thy fame, from granite basins burst,
Leap into life, and, sparkling, woo the thirst?

6

– That thou'rt a proser, even thy birthplace might
  Vouchsafe; – and Mr Cadell MAY, God wot,
  Have paid thee many a pound for many a blot, –

Macadam

Cadell's a wayward wight!
Although no Walter, still thou art a Scot,
And I can throw, I think, a little light
Upon some works thou hast written for the town, –
And publish'd, like a Lilliput Unknown!
   'Highways and Byeways', is thy book, no doubt,
     (One whole edition's out,)
     And next, for it is fair
       That Fame,
     Seeing her children, should confess she had 'em; –
'Some passages from the life of Adam Blair', –
     (Blair is a Scottish name)
What are they, but thy own good roads, M'Adam?

7

O! indefatigable labourer
In the paths of men! when thou shalt die, 'twill be
A mark of thy surpassing industry,
     That of the monument which men shall rear
Over thy most inestimable bone,
Thou did'st thy very self lay the first stone! –
Of a right ancient line thou comest, – through
Each crook and turn we trace the unbroken clue,
Until we see thy sire before our eyes, –
Rolling his gravel walks in Paradise!
But he, our great Mac Parent, err'd and ne'er
     Have our walks since been fair!
Yet time, who like the merchant lives on 'Change,
For ever varying, through his varying range,
     Time maketh all things even!
In this strange world, turning beneath high heaven!
   He hath redeem'd the Adams, and contrived, –
     (How are Time's wonders hiv'd!)
   In pity to mankind, and to befriend 'em –
     (Time is above all praise,)
That he, who first did make our evil ways,
Reborn in Scotland, should be first to mend 'em!

IV

The Scotch Invasion

EACH OF J L MCADAM's sons became a general surveyor of
turnpike roads. Loudon, the youngest, was his father's assistant,
'learning by doing', before he was twenty, but neither William
nor James appears to have had any professional training.
William was manager for his father's British Tar Company.
James was apparently a government contractor. 'I was engaged',
he told the Select Committee of 1823, 'in a mercantile concern
principally connected with Government, in the Barrack department, the Ordnance department, and the Royal Military
College, and other departments of the Government.'[1] What
he was supplying he did not say. The McAdams were not
forthcoming about their business activities.

William, who was married in 1798 when he was barely
eighteen to Jane Pickard, daughter of a Captain in the 13th
Light Dragoons (McAdam connections with the army and
navy were numerous), had three sons, of whom William
(1803–1861) and Christopher (1807–after 1847) each followed

their father's occupation. James's son James (1820–1853) did likewise. James Saunders (1803–?), another grandson of J L McAdam, was on the roads at Bristol (see p. ····) and elsewhere. Col Shaw and his cousin, George Pearson, were road surveyors, if only briefly, in Scotland. Jane, daughter of Christopher Kilby McAdam, who was probably related to J L McAdam through his uncle Gilbert, married John Harding, surveyor to the Cheshunt Trust, one of the many trusts for which James McAdam acted as General Surveyor. Altogether, then, within the McAdam kindred eleven individuals of three generations (see Appendix One, p. 213) are known to have been surveyors of turnpike roads between 1816, when J L McAdam took up his appointment at Bristol, and 1861, when William Jr died, still in office, at Bath.[2]

This was a massive family alliance, far from universally popular. Fine vintages of acrimony between the McAdams and disgruntled trustees, displaced surveyors and professional rivals were laid down year by year and stored in the minutes of turnpike trusts and in parliamentary papers. William cited hostility to his father's activities as his main reason for abandoning his career in Scotland: a move which the Select Committee of 1823 found surprising when he maintained that money was not the main motive.

'At the time my father's system began to expand very much,' said William, 'I felt satisfied, from the opposition it met with in most quarters, that unless my father's sons came forward and supported him zealously, he would not be able to carry the system forward himself; and when I came up to England for that purpose, I did not at all know what the issue might be to myself as to profit, but I came determined at all hazards, to carry on the system for my father's sake.'[3]

James likewise denied mercenary motives:

'I was induced to relinquish the whole of my mercantile concerns for the purpose of introducing my father's system of road-making generally throughout the country . . . I gave up a certainty for an uncertainty, and relinquished a business which I had been fourteen years hard at work in establishing' – meaning that he must have gone into business before he was eighteen – 'and which, at the time I left it, was a very profitable business.'[4]

These passages have a fine evangelical ring about them – 'Go ye into all the world, and preach the gospel unto every creature'[5] – and in that spirit father and sons approached their mission. 'I gave up a certainty for an uncertainty', said James. 'I came determined . . . to carry on the system for my father's sake . . .,' said William, and went on:

'I received my father's direction in the year 1818' – but J L McAdam said he sent for William in March 1817 (above p. 60) – 'when I first came up from Scotland, neither to think of his interest or my own, but to give advice to every trust that asked me, and to teach his system to all persons sent to me, either by turnpike roads or parishes, gratuitously; a number of persons were sent to me, and were so instructed, all whoever applied.'[6]

James took over a surveyorship at Epsom, with 21 miles of road, in December 1817, his father having been too busy to take it on himself. In 1818 William made a start at Devizes, where the local trust had about twenty miles. Loudon probably started earlier than either, when his father sought his help at Bristol.

J L McAdam was rising to the height of his fame, and as a consequence the family was rapidly extending its influence, between 1818 and 1823, especially between 1819 and 1822. Table I shows that by 1823 Loudon had been appointed to sixteen surveyorships, William to nineteen (apart from new roads) and James (Table II) to forty, thirty of them between 1819 and 1822. After the mid-twenties, for reasons which are in some cases apparent and in others can only be guessed at, William and Loudon lost more appointments than they gained. James on the other hand, always the greatest pluralist, gained more than he lost, including the grandest appointment held by any of the family : the surveyorship to the Commissioners for Metropolis Turnpike Roads North of the Thames, responsible for 131 of the busiest miles in the world. The most remarkable expansion, however, considering his age, was by J L McAdam himself. Between the ages of 69 and 77 he added seventeen surveyorships to the one he already held at Bristol : one at Bath, one at Sedbergh, two in Glamorgan, three in Lancashire, four in Cumberland, six in the neighbourhood of Perth. The last surveyorship of his life, on the Perth and Blairgowrie road, he accepted in 1833.[7]

His colleague in the North was John McConnell, much younger and no doubt more continuously active. McAdam's main work for these trusts was probably his first advice, not continuous attention. Two agreements with Perthshire trusts provide for a specified number of visits each year – four, in one case increased later to six – but they make it clear that McAdam was not bound to come every time : McConnell would do. In his last ten years or so McAdam was in the habit of passing the summer in Scotland, and visits to various trusts could fairly easily have been fitted into his plans.

McAdam's experience with Scottish trusts was not wholly successful. In 1832 he and McConnell were involved in a quarrel over a proposal to appoint them surveyors to the North Queensferry and Perth Trust, responsible for about forty miles of the Great North Road, in place of a local man, and after complaints about the condition of the road and the cost of upkeep a motion to dismiss McAdam and McConnell was considered within five weeks of the old man's death. It may be of wider significance that on at least two of the Perthshire trusts and in Lanarkshire McAdam was in rivalry with Telford or his assistant Joseph Mitchell. Scotland was very much Telford territory – he built 1,000 miles or so of roads in the Highlands and elsewhere – and his strength in Scotland no doubt explains McAdam's failure to get himself widely established there.

More surprising, perhaps, than the extensive employment of the ageing McAdam by turnpike trusts was the task he was given by the Prime Minister in August 1826. Lord Liverpool asked him to go to Lancashire 'to endeavour to find some employment for the population'. Distress among the 'manufacturers' was very severe, and Liverpool evidently hoped that McAdam would be able to get them to work on the turnpike roads which, as McAdam observed in 1830, 'were then, as they still are, in bad condition'. When he got to Lancashire and held parish meetings, however, he 'very soon found that no stranger would be admitted to see and intermeddle with the management of the roads and the distribution of the funds applicable to them'.

McAdam therefore had to turn his attention elsewhere and he 'found that many sources of employment remained, such as draining . . . making dams and reservoirs on the streams for turning mills, making alterations and widening *Parish* roads, with many other operations . . . which were successfully resorted to'. He had at his disposal money subscribed for the relief of distress which was being used in aid of the poor rates and 'regularly *given* to unemployed persons'. He immediately put a stop to that, 'as tending to much evil by demoralizing the people and encouraging an opinion that means could be found to give them maintenance without labour'. Instead, he 'resolved to encourage individuals to give work, by offering out of the funds of the Relief Committee, one quarter of the expense of any operation, by hand labour, that the individual wished to have performed; leaving the whole benefit to such individual'.

Under this scheme, McAdam believed, the Committee paid out about £40,000, attracting £120,000 from individuals, making 'about One Hundred and Sixty Thousand pounds paid to hand labourers during one winter, by which a great deal of distress was relieved and considerable permanent improvement made in the county'.[8] McAdam offered the scheme to the Duke of Richmond, a member of Lord Grey's ministry, as a possible method of relieving distress threatening the southern counties in the winter of 1830.

Between J L McAdam's appointment at Bristol in 1816 and the death of his grandson William in 1861, at least 137 trusts in England and Wales and eight in Scotland employed members of the McAdam family as surveyors, and as such they were responsible, at one time or another, for well over 3,300 miles of turnpike roads – about 13 per cent of the total – in England and Wales and for about 400 miles in Scotland.* With some of these trusts they did not stay long, but with the majority, especially among James's, once a McAdam was appointed he remained for life and was frequently succeeded by his son. At Bristol one or both of the two John Loudons were continuously in office from 1816 until John Loudon Jr died in 1857.

Besides the trusts which employed one or other of the family, a large number adopted the system or some version of it, either following advice given personally or in print, or employing a McAdam-trained surveyor. In 1820, for instance, the surveyor of the White Cross turnpike in Yorkshire was instructed to follow 'Mr McAdam's Book on the Repair of Highways' and in 1822 the trustees of the Cambridge and Ely turnpike accepted an offer by J L McAdam 'to send a Surveyor to take the Management of all the Roads in the Neighbourhood at the Salary of 2 guineas a Week he finding himself a Horse'. In 1823 J L McAdam prepared a list of '38 Trusts' – there were in fact 37 – 'none of which are in charge of Mr McAdam's family but repaired by strangers according to his system'. On one of these – Bath – the surveyor in office would probably have denied that he was following McAdam's system, but in 1826 it passed into McAdam's own hands. Of the remainder, half a dozen or so had been or were to be briefly under the care of William or Loudon. The group as a whole, extending from Edinburgh to the extreme South of England, may perhaps be taken to represent McAdam's informal empire, and there were no doubt many other trusts which followed his methods, more or less, without telling him anything about it. It seems safe to conclude that no one else, in McAdam's time or since, has made a more widespread impression, directly or indirectly, upon the roads of England.[9] Telford was a formidable competitor, but his influence was chiefly along the line of the Holyhead road.

* See Appendix One, p. 209.

The Scotch Invasion

If the McAdam surveyorships are plotted on a map, it begins to seem probable that they were not accepted at random. J L McAdam, based in Bristol, badly wanted the neighbouring post at Bath and eventually got it. In Lancashire two of his trusts covered roads in the form of a cross with its centre near Bacup and its extremities at Burnley, Todmorden, Rochdale and Haslingden, with the Liverpool-Prescott-Ashton road not far away. In or near the Lakes there was another fairly coherent group stretching roughly from Milnthorpe to Appleby, and an outlier at Whitehaven. Five of his Scottish trusts centred on Perth and the sixth was the Perth and North Queensferry road. Loudon's appointments, many of which did not last very long, lay in a wide belt of country from the North Riding to the marches of Wales and south as far as Somerset. William set up house at Wilton near Salisbury and concentrated on the country westward from Hampshire into Wiltshire, Dorset, Somerset, Devon and Cornwall, and his two sons worked in parts of the same territory. James's trusts stretched out towards Loudon in the Midlands and towards William in the West, but he rapidly established his main base in and around London, with outposts, some of them briefly held, in the East Midlands, East Anglia and Lincolnshire, and on the other side southward into Kent. Widespread networks of road, stretching through the areas of neighbouring trusts, were thus brought together under the same or very similar methods of management. The most impressive and the longest lasting of these areas of uniform administration was set up by James McAdam between 1818 and 1826, during which time he was appointed surveyor to forty-two trusts with about 1,035 miles of road. Later he lost some and gained others, but the central core of his power base remained intact for the rest of his life : that is, until 1853.

James crowned his collection of surveyorships in 1826 when he was appointed surveyor to the Commissioners of Metropolis Roads, giving him charge of many important streets and of the approaches to the chief roads, North of the Thames, radiating from London :

The Bath Road (A4) by Knightsbridge then through Kensington, Brentford and Hounslow;

The Oxford Road (A40) by Oxford Street then through Acton, Southall and Hillingdon;

The Birmingham Road (A41) by Oxford Street then through Kilburn and Edgware;

The Holyhead Road and the Great North Road (A1) by St John's Street then through Islington and Highgate;

The Old North Road from Shoreditch then through Stamford Hill (A1010);

The road to Essex and East Anglia (A104) from Clapton across the Lea to Snaresbrook.

Each of these routes led, directly or after an interval, to roads under the management of trusts to which James was already surveyor, including for example:

Staines, Egham on the Bath Road;

Beaconsfield, Stokenchurch on the Oxford Road;

Sparrows Herne (through Watford, Berkhamstead and Tring) on the Birmingham Road;

Whetstone on the Holyhead Road and the Great North Road;

St Albans, following immediately after Whetstone, on the Holyhead Road;

Stamford Hill, consolidated into the Metropolis Roads, Cheshunt, Wadesmill and Royston, giving a continuous run of about 70 miles, on the Old North Road;

Hockerill (Bishops Stortford), Epping and Ongar (1830 onward) on the Norwich Road (A11).

Further away from London to the North and East James seemed at one time to be establishing himself strongly with appointments in Northamptonshire, Lincolnshire and Norfolk, but these had all ceased by 1834. He remained strong on the northern mail routes closer in to London. On the Great North Road, taking over at Welwyn in January 1820, he observed: ' . . . this district should rather be considered as part of a more extended Line than as the subject of a distinct and independent appointment.'[10] Later the same year he became surveyor to the

Whetstone Trust on the same road and much later – 1845 and 1846 – added the Stevenage and Biggleswade Trust and the Enfield Chase (Galley Corner) Trust to his appointments along it, giving him a commanding position for twenty or thirty miles northward away from London. On the Old North Road a branch near Puckeridge – the present B1368 – launched him towards Cambridge and in 1839 he took over the Hauxton and Dunsbridge Trust's roads which completed the Cambridge connection both along B1368 and along what is now A10 between Cambridge and Royston. His appointment with the Hatfield and Reading Trust gave him charge of 56 miles of roads between those two towns.

South of the Thames the Epsom and Ewell Trust, the first he worked for, kept up 23 miles in Surrey, including a stretch of one of the roads to Brighton. In 1822 he took over contiguous surveyorships at Sevenoaks, Tonbridge and Wrotham Heath in Kent, to which he added an appointment on the road from Kippings Cross to Flimwell before 1834 and on the road between Chatham and Canterbury in 1837. In 1836 he took over the roads of the Surrey and Sussex Trust, with 61 miles of roads in the neighbourhood of the Borough in South London and a five-figure income. Taken together with his surveyorship to the Metropolis Roads Commission, it brought him as close as he ever came to the comprehensive oversight of the London turnpike roads which he greatly desired. Of the other two brothers, Loudon, in 1823, seemed to have the makings of strength in several districts. Besides being associated with his father at Bristol he had appointments with trusts at Ledbury, Bromyard and Hundred House (the name of an inn some seven miles south-west of Stourport) and he was surveyor also for part of the Kidderminster Trust's roads and part of Ludlow's. Further North he had an appointment at Wellington, another at Chester, and three in North Wales, at Flint, Holywell and Mostyn. He had an appointment on a new road from Buxton to Macclesfield and on new roads out of Worksop to Retford and Mansfield. He was surveyor on roads between York and Scarborough and York and Northallerton, at Pontefract, and to the Stretford Trust near Manchester. Then in 1825 his position was breaking up, for he had lost six trusts – his central three at Ledbury, Bromyard, Hundred House, and besides those, Chester, Pontefract and Buxton – and gained only three : Cheltenham, Lichfield, Shepton Mallet. Only Cheltenham lasted, and by 1835 he had a scattered collection of eight appointments in Nottinghamshire, Shropshire, Carmarthen and Pembrokeshire which can hardly have lent itself to efficient administration.

William, early in his career, acted for his father at Reading. At Newbury, by his own account, he set the Bath road to rights – 'this was formerly the worst Road between Bath and London', he said in 1825, 'has been, since 1819, the best'. He began his career as a surveyor, acting for his father, at Devizes, and in the early 1820s seemed in a fair way to build a block of appointments, much on James's model, with Devizes as a base, stretching to Everleigh on Salisbury Plain, round through Salisbury and Market Lavington, and to Warminster and Melksham in a westerly direction. Much of this holding collapsed, under circumstances to be glanced at later (see p. 93), but William established himself permanently at the Black Dog Trust and at Frome, nearby in Somerset. Black Dog, named after an inn where the trustees used to meet, had charge of the road from Warminster to Bath, and in 1826 when J L McAdam picked up Bath, the whole length of the road came into McAdam hands.

Besides his appointments in Somerset and Wiltshire, William in the twenties was employed at Andover, where he resigned and was then recalled – 'since then everything has gone on with regularity, and the Roads are getting into excellent order'[11] – round Winchester and Southampton, and in Dorset. In Devon he became very strong, and his work there is discussed in Chapter VI. In Cornwall he was employed at Launceston and Truro, and his son Christopher, besides working with his father at Truro and Exeter, was employed independently at Launceston and other places in Cornwall and Devon.

William in the West of England seems to have made a

speciality of working for limited periods on new lines of road sanctioned by special Acts, as Loudon evidently did in Yorkshire and the Midlands. At Cerne Abbas in Dorset, where a new trust was established in 1825, William and one of his sons, probably the younger William, were employed for two years at a fee of £300, though here there seems to have been little in the way of new construction. By Michaelmas 1825 he had received nothing. In Somerset and Devon there was a good deal of new building. In 1825 William reported to Parliament work on new roads at Minehead (14 miles), Haldon (4½ miles), Taphouse (6 miles), Torbay (3 miles) and Modbury (14 miles). The roads at Haldon and Taphouse were built for the Exeter Trust, to which William was surveyor: the Torbay and Modbury roads perhaps for two other trusts which employed him – Totnes and Plymouth.[12]

If the McAdams were called in, it was usually because a trust found itself in difficulties, often after a committee of trustees had examined the finances and the state of the roads. The first approach would probably be to J L McAdam, who might or might not report personally. Then, unless he was proposing to take the post himself, he would send one of his sons, who would inspect the roads and report, if his father had not done so, and make suggestions, usually with a cheering estimate of how much money would be saved if they were followed. The trustees would then take him at his word and appoint him, either dismissing the surveyor in office or ordering him to work under the newcomer.

The way the process worked is well illustrated in correspondence preserved with the trustees' minutes of the Truro roads, who in 1825 found themselves under the necessity of finding a surveyor to replace their Mr Hayward who was resigning, perfectly amicably, because his health was giving way. The Trust was responsible for about 45 miles of roads radiating eight or nine miles from Truro. The bonded debt was £3,600 and the income, the tolls having recently been increased by 25 per cent, was about £2,200.[13]

Humphrey Willyams, one of the trustees, was deputed in

April 1825 to write to J L McAdam, who promptly sent the
letter on to William, remarking 'he will be able probably to
put you in a way of having your Roads managed better and with
economy'. William met the trustees and in June wrote to them
explaining in detail how 'we' – presumably himself and his
sons – managed the affairs of trusts employing them:

'Our mode of proceeding is to take charge of a Trust and
expend on it any sum of Money the Trustees can afford to
lay out in ordinary repairs, drawing it in advance in 13 lunar
monthly Installments of which we render a monthly account
with Vouchers, paying every thing in ready money. Your
Trust is about 50 miles in Length and your expenditure
about £2,000 a year and as your Bonded Debt is not large –
your revenue I should consider sufficient to entitle you to
excellent Roads . . . I should place a resident Sub Surveyor
on the Trust and allow him a *sufficient* Salary for himself
(& to keep a horse if the lengths of the Branches make it
necessary) but *that* is of no consequence to the Trustees for
he is paid out of whatever sum we are allowed for the repair
of the Roads. We consider such a Man as the sine qua non
of a good Road, as much as the Materials laid on it and if it
produces a good Road that is enough. I should not object to
take charge of the Trust for a period of 5 years: but I would
not engage unless for a *fixed period* because the Expence of
reforming the Roads finding proper Quarries &c would
probably leave me little or no recompense for the first 2 years.
But as it would be the duty of the Trustees to see that the
public Roads are attended to, the Trust ought to be at
liberty . . . to name 5 . . . of their own body as a Committee to
inspect the Roads . . . after one year and if a Majority (the
Chairman having a casting vote) are not satisfied, hearing
my explanation, . . . then to end the bargain on 6 months'
notice. This would secure me from unjust opinions of
persons not liking my being called in . . . My Salary would
depend a good deal on the traffic and consequently inspection
and care necessary, also in some degree on what the Trustees
could afford and were justified in paying . . .'

As a less eligible alternative, he suggested 'inspecting and reporting on the *Roads* only'. If they preferred that, he would do it, 'but to do so *effectually* I should be obliged to charge you at least one year's Salary if not more and you would have no security that you got anything for your Money as I should not be responsible were anyone employed to carry my views into effect to tell you what I said was worth 2/6 had cost 5s.' He backed up this somewhat incoherent warning with 'a case in point' from the Plymouth roads. We sent a man there', he said, 'of whom we had a good opinion [but] he found himself without a Master to controul him, and handling Money, became a Blackguard and was sent away.' The Plymouth trustees then sent a man for William to teach, but that was no better. 'After working on for 2 or 3 years at what he called our plan I was call'd in and at our Audit last Feby at the end of the second year . . . my Expenditure in the *two* years was as nearly as could be the amount of the *one* year preceding my taking charge.'

William warned the Truro trustees against putting their repairs out to contract. 'Whenever a Trust *apply* to us', he said, 'we give them all the Information we can and if employed do them all the good in our power, our Salary once fixed our Interest becomes that of the Trust. First in having good Roads, next in having cheap Roads. What the Trust gain in funds we gain in credit. Whereas a *Contract once signed, the Contractors Interest and that of the Trust become diametrically opposite – what one gains the other loses.'*

Impressed, evidently, by this powerful presentation of the McAdam System, Willyams explained on behalf of the trustees that the Truro roads 'tho' in themselves in a very good state of repair are kept at too considerable Expence – and our object is by a better system to save sufficient to expend in an alteration and Improvement of our *line* of Road; which from its partial steepness and narrowness, is very defective.' That there was 'a defect somewhere in the management,' he added, with a slightly puzzled air, 'the Trustees are all agreed, but on the remedy they are divided, and to you they refer for your advice.' What salary would Mr McAdam require under the terms of a five-year agreement?

William replied that he would expect £150 a year. He was receiving, he said, £210 at Plymouth, £250 at Totnes, £220 at Frome – 'and these sums may appear considerable yet when diminished of *all* my Expences does not leave me a great deal to carry to my Family in lieu of my time and trouble.' If at the end of any year the trustees 'shall vote that I don't deserve my Salary,' he said, 'then I will go without it . . . for I never have and never desire to receive a shilling from Trustees unless accompanied by . . . their approbation and goodwill.'

He thought that perhaps 'for want of a proper system' the Truro trustees were paying too much for their work. 'I should never have reduced prices,' he went on, 'had I not known how to show the Men the way to make a little more Money at lower prices. Laborers are ignorant themselves and require patient instruction. I am generally as unpopular at first as I am the reverse at last with them.'

Finally he came to his proposition:

'Suppose I were upon inspection of your Trust to think I could undertake it for five years *thus*, how would it answer your objects? (and if I did so it must be distinctly understood that I will not let the Road go for ruin or get into a failure myself, if I see after a little experience difficulties exist not to be surmounted but at a greater outlay I will tell you so and be perfectly ready to relieve you of the weight of my personal charge if required).

Gross Revenue	£2,200
41 miles @ £31 including Sub-Surveyors' Salary	£1,400
Gen. Surveyor's Salary	150
Interest on Bonded Debt of £3,600 at 4 per cent	144
	£1,694
b/f	£1,694
Other Salaries repairs of Tollgates & houses & other incidents – say –	156
Sinking or accumulating fund say	350
	£2,200'

Watering a road, 1805. There are many references to road watering, to keep down dust in dry weather, in the records of the Metropolis Roads' Commission in Sir James McAdam's time (1827–1851).

Fotomas Index

The Stone Breaker, 1858. This boy seems to have learnt McAdam's lessons well. He is sitting, using a short-handled hammer, and breaking the flint small, though perhaps not small enough.

Fotomas Index

One of the gates belonging to the Bath Turnpike Trust, on the
London Road, 1826.

Bath Reference Library

A stage-coach at a turnpike gate, evidently on a wet evening.

Fotomas Index

The trustees were convinced. They fixed a date, at William's convenience, for him to inspect the roads, and on 18th June 1825 voted him into office as surveyor. The whole negotiation had taken a little under two months.

On his way to inspect the Truro roads, about 12th June, William called in at Launceston to report on the state of the roads there. On 19th July the trustees decided to appoint him in place of their existing surveyor and at a higher salary – £125 instead of £70. Four years later they were proposing to pay part of his salary by way of mortgage bonds – always an occupational hazard for turnpike road surveyors – but they thought better of it at their next meeting and found the cash.[14]

By the time he made his agreements with the trustees at Truro and Launceston, William McAdam had been working in England for about seven years, and however sincere he may have been in saying that when he left Scotland he was not concerned about material rewards, it is evident that by this time he was determined, quite justifiably, to see that trustees paid him a proper rate for his job. He obviously thought of himself as a partner in a family firm, and among the services offered by the firm was the provision of sub-surveyors whom he, his father and his brothers looked upon as their own employees to be placed with one trust or another as McAdam, not the trustees, might think fit. J L McAdam said in 1825 that there were about 100 sub-surveyors thus employed. Trustees, naturally, did not always accept the McAdam view, especially when one of the family proposed to move a sub-surveyor whose services the trustees valued. Disagreements arose, and the family did not always win.

Where the process of appointment to surveyorships can be followed in detail, it is often possible to trace the influence of some great man – the Earl of Hardwicke, Earl De La Warr, the Earl of Verulam – behind it – and, as a great landowner usually had connections with every trust in the neighbourhood of his estates, employment by one trust might be followed by employment by others adjoining or nearby. This was very much how James McAdam came to accumulate so many appointments round London, and the way it was done can be watched in the minutes of three trusts – Cheshunt, Kneesworth and Caxton (or Royston), Wadesmill – between August and December 1818.

The main responsibility of the Cheshunt Trust was a stretch of a dozen miles or so (now A1010 and A10(T)) of the Old North Road, running from the boundary between Enfield and Cheshunt through Ware to Wadesmill and forming part of the mail route to Edinburgh. In the late summer of 1818 the trustees commissioned a report from James McAdam which was read at their meeting on 21st September. On the strength of the report they appointed James Surveyor from 10th October 1818 and dismissed the surveyors then in office. Having put in their energetic young Scotsman, the Trustees cautiously waited for over a year before voting him any money. Then on 11th November 1819, with their Treasurer, the Revd Mr W A Armstrong, in the Chair, they confirmed a proposal already made to pay him £250 'for his past Services' and made another proposal for 'a permanant Salary of 150 Guineas per Annum . . . to commence from the 10th October last [ie: 10th October 1819]'.

They accompanied all this munificence with a tribute which James may have received with mixed feelings:

'The Trust in awarding to Mr McAdam £250 as a proof of their satisfaction of the very great Benefit which the Roads under their superintendance and management have derived from the introduction of his excellent System of making and repairing Public Roads and of giving employment to the Poor in the different Parishes, is perfectly sensible that the sum is a very inadequate remuneration for his Services, and is certainly less than the present good State of the Finances of the Trust might afford, but the Trustees as such, have not felt themselves justified in indulging in feelings of liberality at the expence of the Public, and they are likewise aware that the award of a larger sum might as a precedent have detored [sic] other Trusts in less flourishing circumstances from availing themselves of the benefit of his Services.'

They went on to thank 'Mr McAdam senr' – to whom they had probably applied for advice in the first place – 'for his occasional inspection of the improvement made on the Roads of the Trust' and directed that both he and James should receive a copy of 'these Resolutions', which were also to be published in the *County Chronicle*, the *Farmers' Journal* and the *New Times*. This was free publicity for the McAdams, no doubt welcome, and also a plain indication to other Trustees that they need not be frightened of the expense of the McAdams' services.[15] It is to be hoped that James and his father were both well pleased.

North of the Cheshunt Trust's territory the Wadesmill Trust took over as far as Royston, and thence for about forty miles, to Wansford a few miles short of Stamford, the mails ran along the Old North Turnpike Road, once Ermine Street and now A14(T), and a few miles of the Great North Road (A1(T)), managed by the Kneesworth and Caxton Trust.[16] The Kneesworth trustees became dissatisfied with their roads and their surveyor about the same time as the Cheshunt trustees were having doubts about theirs, and their answer to their problem was the same: send for McAdam. Probably the suggestion came from the third Earl of Hardwicke (1757–1834), one of the trustees and an enthusiast for macadamization. The road was administered in three divisions, of which the southernmost, running for about 15 miles out of Royston, passed the Earl's park gates at Nuneham Wimpole. On 1st October 1818, James was asked to report on this fifteen-mile stretch to a committee of trustees which included both Hardwicke himself and another earl – De La Warr.

A little over a month later, other work notwithstanding, James put in his report, dated 3rd November. It is a typical McAdam document: brief, pointed, self-confident. 'The first six Miles', he said, 'being made of naturally small-sized Materials are Comparatively smooth, but require to be properly formed for the immediate discharge of the Water, and are very much loaded with Sand, and dirt, and much cut into Ruts, which would not occur were the Materials properly prepared in the [gravel] Pits and due attention paid after being laid upon the Road until it became perfectly smooth, hard, and solid.'

After that sharp jab at the previous management he turned to the remaining 9 miles of road which he found in the state which had so often incensed his father – 'much overloaded with Materials which have been thrown upon the Centre of the Road in a very improper state and have been allowed to arrange themselves in the Rows in which the Wheels of Carriages passing through not over these large Materials may have placed them, rendering the Road very Rough, full of deep Ruts, and at many places even dangerous.'

Just as the malady was described exactly in his father's terms, so the remedy was exactly what his father would have prescribed, and it may be that James had his father's help in preparing his report. 'Were the Materials', he said, 'of which this part of the Road is composed properly separated prepared and relaid, due attention being paid to the formation of the Road until it became perfectly smooth, hard and Solid, I am of opinion that few or no additional Materials would be required on the greater part for a length of time and upon those parts that might require to be strengthened a very small proportion of the Materials properly sorted prepared and applied would form a firm smooth solid Road equally good at all seasons of the Year.'

He went on to ram home the argument which all trustees found most telling – 'think of the money you'll save'. 'By the adoption of this System,' he said, 'the Expenditure of such Vast quantities of Materials would Cease; altogether with the great expence of obtaining and Carting them, the Operations on the Road would be almost entirely confined to Manual Labour affording employment for the Labourers of the respective Parishes through which the Road passes' – another shrewd point, aimed at trustees who were also ratepayers – and he ended: 'I beg respectfully to recommend to the consideration of the Commissioners [Trustees] the expediency of letting their Tolls [instead of keeping them in hand, presumably], by which measure and the Roads being put into good and proper

Condition, I have no doubt the Revenue of the Trust would be considerably improved.'

Faced with this strong pitch for their business, the trustees called a Special General Meeting at the Hardwicke Arms Inn, Arrington (near Wimpole Hall), which the two Earls and nine other trustees attended. They put to James two questions, both of which, with his answers, they recorded in their minutes. He had evidently estimated the cost of the work needed on the road at £1,035, and they asked him whether, if that sum were to be spent 'by Mr McAdam, in carrying his system into effect', he thought the Trustees would also have to spend 'the same sum which they have been hitherto accustomed to apply to the Ordinary repair of the Road; or by placing the whole responsibility in the Hands of Mr M'Adam will the ordinary expenditure of the Road be included in that which he has calculated to be necessary to put it in a good state of Repair?' His reply was bold to the point of insolence : 'I mean that if that sum is expended in the course of the next six months, the road should be at that time in a good state of repair – with respect to the latter part of the Question, it is impossible that there can be any other expenditure; because the two operations would interfere with each other.'

Nevertheless the trustees, evidently determined to get James to commit himself irrevocably to carrying out his promises, pressed him a little further, to be received once again with self-assurance tinged with a hint of scorn :

'Q : Does M'Adam conceive that after the road shall have been put into a proper and substantial state of repair with the above sum of £1035, according to the system he has adopted; the future expense of maintaining and keeping it up, will be greater or less than it has hitherto been?

A : I should suppose that after the Road shall have been put into a good state of repair the expense of repair will be infinitely reduced.'

With the prospect of infinitely reduced expenditure thus so confidently held out, and recorded in writing, the trustees engaged 'M'Adam' (they consistently refused him his small c) 'to undertake the putting of the Road into a Complete state of repair according to the Estimate.' They passed on to arrange a loan at 5 per cent, guaranteed by the two earls, from Messrs Fordham & Co of Royston – Edward King Fordham, present at the meeting, was Treasurer of the Trust from 1803 to 1824 – and they rounded off their business by appointing Mr M'Adam 'Surveyor of this Road during the pleasure of the Trustees.'

The trustees thus decided to back James with £1,035 to carry out his recommendations and produce for them the benefits so confidently forecast. They made no commitment to pay him anything at all, for neither fee nor salary is mentioned in the minutes, though there was clearly a moral obligation to pay for success and verbal promises may have been made. James, in any case, had the strongest possible incentive to make good his undertakings as fully and as rapidly as he could. It was more than a year after his appointment before any money was offered at all, and when the offer was made, in January 1820, it was couched in phrases strongly reminiscent of the tribute published by the Cheshunt trustees in November 1819 :

'. . . it appears to the Trustees at this Meeting [6th January 1820] that Mr M'Adam has rendered and is rendering essential Service to the Trustees in the improved Condition of the Road under his Care and the superior economy of his System. That Mr M'Adam be invited to continue his Service as General Surveyor during another Year. That the Trustees express to Mr M'Adam their regret that the state of their finances does not allow them to remunerate him adequately to the Benefit they have derived from the Application of His System and that the sum of One Hundred Pounds be now presented him for his Services in the present Year – And that the Chairman be requested to Communicate these Resolutions to Mr M'Adam.'

The next year matters were worse. 'The Trustees,' on 4th January 1821, 'regret that in offering to Mr McAdam the sum of Fifty Pounds for his services in the past Year . . . their

Remuneration should be limitted by the State of their Finances.' Mr McAdam might have reflected with some bitterness, though there is no evidence that he did, that the Cheshunt trustees, who could pay well, would not, out of regard for 'Trusts in less flourishing circumstances', and that the trustees at Kneesworth and Caxton had shown no hesitation in taking the hint thus dropped.[17]

Perhaps he found it all worthwhile. On 3rd December 1818, Lord Hardwicke became a trustee of the Wadesmill Trust, which had charge of the 28 miles of Old North Road connecting the two trusts, Cheshunt and Kneesworth and Caxton, to which he had already been appointed Surveyor. In any case, by what looks like stage management James came to be waiting in the wings when Lord Hardwicke took up his trusteeship, for another trustee, Daniel Giles of Youngsbury, Wadesmill, 'stated that Mr James M'Adam who hath the Management of the Cheshunt and Kneesworth Trust was in Attendance and desirous to communicate with the Trustees upon the State of this Road'.

He communicated and was asked for a report, which he put in on 16th December. It was in very similar terms to the report he had recently made for Kneesworth, and indeed in some passages almost identical in its wording, except that it included a plea for a 'Skilfull, active and Trustworthy Sub-Surveyor,' who 'should exercise no other occupation'. At a meeting on the following day, 17th December, with Hardwicke in the Chair, the report was accepted and the road was immediately put under Mr McAdam's care for one year. He served what seems to have been the customary year's probation without pay and then on 7th January 1821, again in Hardwicke's presence, announced that he had received £2,866 1s 5d (£2,866.07) and spent £2,856 6s 4d (£2,856.32). The trustees were very pleased. They voted him £250 for his year's work and appointed him to 'the Survey and Management of the Road' at £150 a year.[18]

As surveyor to the Cheshunt, Wadesmill and Kneesworth Trusts, James McAdam was in charge of the Old North Road along the whole of its length from Cheshunt until it joined the Great North Road at Alconbury. It is difficult to believe that the neat dovetailing of these three appointments in the Autumn of 1818 came about by chance. It seems almost certain that Hardwicke was determined to see McAdam in charge of as much of the Old North Road as possible, and used his widespread influence accordingly.

Hardwicke also sponsored James for the surveyorship to the St Albans (Pondyards and Barnet) Trust which had charge of 11 miles of the Holyhead Road : an even more important route out of London than the Old North Road, and the especial concern of a body of Parliamentary Commissioners. The Trust's road ran from Hadley Pillar just North of Barnet, which still stands, commemorating the Battle of Barnet, 1471. It ran roughly along the present line of A1081 and A6(T) to Barnet and by a complicated route right through the town – unusual for a turnpike road – to Pondyards just short of Redbourn on the other side.

Early in January 1820, a report on the St Albans road caused the trustees 'no inconsiderable degree of alarm'. They resolved 'to take into consideration the present involved State of the Finances of the Trust and to enquire into the Cause thereof; also to consider the general dilapidated state of the Road . . . and whether some Measures cannot be adopted for keeping the Road in a better state and at a much less annual Expence than £170 and upwards per Mile, which is the actual Expence the Trustees have been put to during the last Year, exclusive of Interest Money and Contingencies.'

A little over three weeks later the trustees decided 'that the present involved State of the Trust Funds is owing to the improvident and unecessary Expence which the Surveyor has put the Trust to in the Repair and Management of the Road and to his general Incompetency to the discharge of his Duty.' A resolution to dismiss him was then passed and the wretched man, Henry Norris, was called into the room to hear it, being later prosecuted for his financial mismanagement.

This was a situation made for McAdam, and as soon as

Norris was out of the room two letters were read, one from Lord Hardwicke and another from Lord Verulam, both large landowners in the neighbourhood of the Trust, recommending James. The Trustees 'unanimously approved of such Recommendation', and James, called to meet them a week later, accepted the post. With his consent, the question of remuneration was deferred for a year. In March 1821, after he had reported that his first year's expenditure had necessarily been heavy, 'the Road requiring to be lifted, formed, Water Courses made, Coated &c.,' his salary was fixed at £120 a year. 'The expenditure of the second year,' he wrote, 'will I trust be considerably less and the future expence will be found to diminish in the exact proportion that the Road is kept in good Order.'[19]

Minutes, revealing though in many ways they are, scarcely convey the full force of the impact of the McAdams upon English turnpike administration. Charged with reforming zeal and self-confidence, not to say arrogance, the family burst upon the trusts with all the delicacy of exploding handgrenades, shattering the peace and perquisites of trustees, surveyors, clerks, treasurers and contractors all alike. That they brought with them unquestionably superior methods of road management naturally pleased their sponsors, and great men like the Earl of Hardwicke could take a fairly detached view, but it did nothing to soothe the indignation of surveyors turned out of a comfortable billet, trustees deprived of local patronage, clerks and treasurers jolted out of an accustomed routine, and the fact that the incomers were Scotch (in the usage of the day) made their offence much worse in the eyes of the injured English natives.

William McAdam cited opposition to his father's methods as the reason why he came to his father's assistance (see p. 72), and he ran into hostility himself in a group of Wiltshire trusts – at Devizes, Salisbury, Warminster, Melksham and Westbury – from which he either resigned or was dismissed. At Devizes, his first appointment, the trouble arose, according to his own account, because the trustees induced his father to retain the old surveyor. 'His opposition to my arrangements', said

William, 'was so great, that I had to send for my nephew to assist me, and we had to do the sub-surveyor's work also for twelve months nearly; and after I had forced the new arrangements to completion, the sub-surveyor worked so extravagantly, I reported him, but was given to understand, if I meant to remain I must retain him. With my Father's advice I resigned the Road, not choosing to be concerned in such a waste of public money.' That seems to have been in 1820. At Salisbury in 1822, after about four years' work, 'having, by great labour and pains, brought these Roads from a ruinous condition into good order for less money than I was ordered to expend, I was discharged, principally by the endeavours of two Trustees, whose lands I entered for materials by orders from the magistrates, supported by the farmers' trustees who had neglected their statute labour till the Roads were indicted; in which state they were delivered to me.' He gave up the Warminster roads at the same time, though he does not say why.

At Melksham he collided with the clerk who, he believed, instigated 'a few Trustees' to return 'fallacious accounts' to the clerk of the peace in William's name but without his knowledge, and then, in July 1823, dismissed him. 'The clerk', he recorded, probably with satisfaction, 'became treasurer, embezzled the funds of the Trust, was sentenced at the last Wilts assizes to transportation; the Trust affairs are in the utmost confusion . . . I attribute the whole of this confusion to the neglect of duty in the Trustees.' In September 1823 he was dismissed from his post at Westbury, noting acidly : 'Made these Roads good, and found for them new quarries of better materials, during these three years [29 Sep 1820 to 29 Sep 1823], at the end of which I was discharged, and my sub-surveyor retained.'[20]

This string of resignations and dismissals has all the appearance of being the result of an organized anti-McAdam campaign throughout the neighbourhood. If it was, it had influential support. When the question of a grant of public money for J L McAdam was debated in the House of Commons in May 1825, the grant was opposed by T G B Estcourt, High Tory MP for Devizes, over which his marriage to an heiress had given him what the DNB calls 'the chief influence'. Mr McAdam, Estcourt said, had left the roads round Devizes much worse than he found them and – which was probably a greater offence – 'had also thought it proper to asperse the conduct of the trustees of those roads in a manner highly unjust . . .'[21] Just as the McAdams could attract the favour of some great men, they could as well fall foul of others, and the results were equal and opposite.

It is fair to balance Estcourt's attack with the evidence given to the Select Committee of 1823 by the Treasurer to the Exeter Trust, to which William became surveyor in May 1820. McAdam's system, Ellicombe said, brought the roads into 'high condition as to strength and smoothness . . . Several carters have admitted that the roads already afford an advantage nearly equal to one horse in three.' Moreover the system was economical – 'The annual supply of materials has so evidently overtaken the annual wear as to justify the trustees voluntarily . . . to abandon the whole composition in lieu of statute duty, amounting to £705 per annum.' He calculated the average expenditure over seven years, before McAdam's time, at £6,756 8s 9d (£6,756.43), or about £45 a mile. Under McAdam, the average expenditure over three years was £4,959 6s 7d (£4,959.32) – about £33 a mile – showing a saving of £1,797 2s 2d (£1,797.11). Moreover McAdam's system of accounting, with 'minute and daily control over the expenditure', had enabled the trustees to pay off debt at about £1,000 a year, raising the bonds of the trust from 80 to par, 'and must ultimately be a more general bonus to the public by the reduction of tolls.'[22] In other words, a textbook McAdam performance both on the technical and the administrative side : exactly what every trust that employed any of the family hoped to see, and what the family with great assurance undertook to deliver.

Nevertheless, with so many trusts to attend to there was no possibility that William or either of his brothers, having set up their system, could carry out in detail the daily tasks of management, and none made any pretence of doing so. 'I am differently situated to my father,' said William, 'who resides upon one trust,* and enters into all the detail of the accounts; I am scattered over so large a space that I merely lay out such sums as the trustees tell me they can conveniently spare for the repair of the roads, they arranging all the finances . . . and all other necessary expenses, giving me the difference for the repairs.'[23]

The brothers, indeed, always acted, especially in the way they accumulated surveyorships, as if their responsibility lay more towards the public in general than to any particular body of trustees, and James in 1839 said publicly, as his father had said before him, that there were too many trusts, and they ought to be grouped together rather as the parishes, by the time he spoke, had been grouped into 'unions' for the purposes of the new Poor Law.[24] Consistently with this opinion, James and his brothers described themselves as 'general surveyors' and they saw their duty as that outlined for a general surveyor by their father in his *Management of Trusts* (see p. 45): to advise the trustees on the one hand, and on the other to supervise the sub-surveyors on the roads, but not to do their work for them. In this way, if Parliament would not set up the centralized system of road administration advocated by J L McAdam and James, and probably the other sons too, then the McAdam family would come as close as they could to setting up something of the sort themselves.

From scattered entries in the minutes of various trusts it is evident that from time to time the McAdam brothers fell into the same trap as their father at Bristol, and over-extended themselves, particularly perhaps William, against whom there may also have been other grievances. His surviving correspondence suggests a brusque, aggressive personality, perhaps with a touch of acid humour, which served him well in getting employment, when trustees wanted to save money and to see a failing surveyor discredited, but perhaps less well in keeping it, when trustees wanted a compliant and punctilious official.

---

* J L McAdam had not yet entered into the multifarious commitments of his last years.

Furthermore there are indications that he may not have been quite so efficient as he claimed. The Truro trustees more than once grumbled about his presentation of accounts, and both they in 1828, and the Launceston trustees in 1832, disallowed charges which they considered unjustified.[25] By 1833, the Truro trustees, having decided that they must reduce their expenditure for upkeep, seem to have been disenchanted with William. They put an end to the joint agreement which by then they had with him and his son Christopher and replaced it by one with Christopher alone, for three years not five, and at half his father's salary.[26] It is no doubt partly in directions such as these that we must look for some of the reasons why William fell foul of so many trustees. Even James, far more successful in holding appointments, was sometimes sharply reminded of his duties, as when the Kneesworth and Caxton trustees wrote to him in April 1829:

'From the circumstance of your being seldom able to give us the advantage of your personal inspection, a greater degree of responsiblity and a larger share of duty has fallen upon the Sub-Surveyor, and we think it due to Mr Riddeford to State that he . . . is acquainted with the sort of attention as well as the description of material that it [the road] requires.'[27]

In the view of the gentry road surveyors were, after all, no more than upper servants, like other professional men.

Since none of the brothers seems to have found any difficulty in switching at short notice into a road surveyor's post from a totally different occupation, or in teaching others after the switch was made, it can hardly be contended that much was required in the way of specialized training. Indeed, as we have seen, McAdam's technique of laying a road surface required nothing more than close attention to simple details, and on the administrative side, on which he laid equal emphasis, it was similarly a matter of setting up a simple system of contracting and of financial control – and then seeing that it was followed. The job required a sound commercial sense for dealing with contractors and a sharp eye for supervising those who carried out the simple duties required to keep a turnpike trust running and solvent. It also required a sound constitution and tireless energy, for a general surveyor was constantly on the move. As James put it, 'my only home is the turnpike road three or four nights of the week, or three at all events, as well as seven days', and from what little can be gathered of his movements, in trust minutes and elsewhere, it is easy to believe his claim. On Monday, 10th February 1834, for instance, he wrote to the Duke of Richmond at six in the evening to say that he had just returned to Town and was about to leave again 'per Mail' for Grantham, and would be back 'some time on Wednesday'. The York mail coach, which would take him to Grantham, left from the Bull & Mouth in the City at 7.15 pm. He was probably writing from an office near Charing Cross, so he would have had to get to the City in time to pick up the coach and then travel through the winter night until eight next morning, with the return journey to be faced the following day. It sounds like the life of a busy executive, which he was.[28]

The McAdam brothers, as general surveyors, found themselves obliged not only to pay coach fares and for the hire of post-chaises, but also to keep vehicles, horses and their attendants. Moreover they needed professional assistance. Most of the cost of all this had to come from the salaries they received from trusts, though some paid travelling expenses and the larger salaries, such as J L McAdam's from Bristol and Bath, and William's from Exeter, were fixed with expenses in mind.

William, in 1823, was employing his son as an assistant. James, more ambitious, was finding three assistant general surveyors to whom he paid, he said, 'reasonable salaries' and for whom he provided two gigs and horses, for they were 'constantly on the road'. For himself, he said, he travelled in 'every kind of conveyance'. He kept a phaeton, three pairs of horses, and the necessary attendants. William kept a light carriage and two horses in Wiltshire and a gig and a horse in Devonshire, and he travelled by horses, coaches and chaises.

'I have so many miles to go over . . .', he said, 'that I travel all ways'. Loudon, the youngest of the family, employed an inspector, but said nothing about finding his transport.[29]

These expenses cut deeply into salaries and it is difficult to estimate quite what the net income of various members of the family amounted to. Almost certainly one reason why J L McAdam, as an old man, accumulated appointments in the North was that he wanted to add to the sums he was getting from Bristol and Bath and towards the end of his life he must have had an income running well into four figures from which, if he arranged his travelling skilfully, he might expect to have a comfortable balance remaining, but he would have had to share it with McConnell.

William seems to have been plagued by trusts which could not or would not pay. In 1825 he mentioned several which had paid him nothing, others which had failed to pay the sums agreed, and four which forced bonds on him instead of cash. At Westbury, Wilts, the clerk compounded this injury by offering to redeem three bonds of £100 nominal, supposed to represent three years' salary, at 65. Payments for road-building, on the other hand, seem to have been good and reliable – £420 for 4½ miles at Haldon; £500 for 6 miles at Taphouse; £157 10s for 3 miles at Torbay; £500 for 14 miles at Minehead; £700 for the same distance at Modbury, though in both these cases it was a matter of taking something on account and waiting for the rest.[30] In 1828 he agreed to £350 for a new road near Truro, to be increased by £50 for every three months' delay caused by the trustees, who for their part wanted it 'distinctly understood that Mr McAdam shall give as much as he can of his personal superintendence to the work . . . and that when . . . he cannot himself attend, Mr McAdam's son should do so in his stead.'[31] Trustees might have to put up with absentee surveyors, but they did not really like them.

James, who consistently gives the impression of having been the ablest, the most businesslike and the most successful of J L McAdam's sons, early acquired the thoroughly professional custom of charging in guineas, though he seems more often to have had to content himself with pounds. Most of his salaries were in the range £100 to £105 (100 guineas), with only two above (£157 10s – 150 guineas) and a few below.[32] He complained neither of non-payment nor of payment in bonds. Evidently he chose his trusts well.

William, working in equal partnership with his son, computed their total gross cash receipts from Michaelmas 1818 to Michaelmas 1825 at £11,052 12s. To that he added £1,100 in Road Bonds valued at 65 – £715 – and he optimistically added £2,455 of unsettled claims, giving total receipts, actual and notional, of £14,222 12s. Against that he set £7,700 travelling expenses for himself and his son, leaving £6,522 12s for seven years or £911 16s a year yielding £455 18s each* 'for their time, skill and statute labour given up, while excellent Roads were kept, on all their Trusts . . . in summer and in winter. *Provided* they ever do receive, without further expense, the full amount of all unsettled claims.' He does not sound hopeful. Loudon's case was worse. In 1823 he told the Select Committee on his father's petition that his salaries amounted to £630, but by the time he had paid his inspector's salary of £150 and travelling expenses for both of them, he had £180 left. By 1825, having lost six trusts and gained three, his income was £100 less, but he did not say what his expenses were.[33]

James in 1823 had fixed salaries from all his trusts except four, to a total of £3,479. In 1825, he said, 'a diminution' had taken place of £267 per annum 'owing to his health having suffered so much as to prevent his being enabled to attend to the whole of the arduous duties of his office'. Nevertheless he had been working for various authorities in London, no doubt preparing the ground, wittingly or unwittingly, for his appointment as Surveyor of Metropolis Roads in 1826. From the Commissioners of His Majesty's Woods he received £150 for 'making the Albany Road', presumably Albany Street beside Regent's Park, £150 for 'two years service in the several Parks', £100 for 'making new Roads in the Parks'. The

* William's arithmetic let him down. The true figures are £931.80 and £465.

Trustees of St James's Square paid him £40 for 'forming a granite Road, and keeping same in repair', and the Commissioners of His Majesty's Navy £15 for 'forming Rolling-way in Deptford Victualling Yard'. The Commissioners of Pavement at Colchester paid him £16 10s (£16.50) for 'M'Adamizing their streets'.[34]

The very large sum of £3,479 a year attracted comment. James knew it would, and he had his defences ready. It represented only £4 1s 1d (a shade over £4.05) a mile 'for superintendence, including all expenses': a figure which his father compared with £7 2s 9d (just under £7.13) a mile received by a rival surveyor in Bath. James's two assistants, with their travelling, cost him £700–£800 a year, and he estimated his whole expenditure, including the upkeep of a London office, to be 'at least £1,600–£1,800'. Of the £471 10s which he received for work in London and Colchester between 1823 and 1825, he calculated that £260 3s (£260.15) went in expenses, leaving £211.35 spread over a couple of years. In October 1826, when he was appointed Surveyor to the Metropolis Roads, the Commissioners voted him £500 a year plus £200 expenses, and they required him to find security for £5,000. At the peak of his career, therefore, it seems likely that James McAdam's income from surveyorships was probably between £2,000 and £2,500 a year, clear of expenses.[35]

This was in a society in which the main source of wealth was still the ownership of land and middle-class employment of any kind was desperately scarce: how scarce is shown by the fact that there was always competition for lieutenants' appointments in the Navy, and yet for many years after the end of the Napoleonic war about three-quarters of naval lieutenants were unemployed on half pay of 5s to 7s (25p–35p) a day.[36] The McAdams, at the levels of income discussed above, were inventing a well-paid profession. Successful London physicians and barristers would earn more, some much more, than James, but he was well able to look prosperous attorneys and surgeons, as well as businessmen, in the eye. His brother William and his nephew of the same name, with incomes ranging between £450

Macadam

and £600, clear of expenses, had plenty for solid comfort. Loudon in 1823, at about the level of a lieutenant's half-pay, was much less well off, but in 1823 he was still a young man with his way to make : his position was far from intolerable.

As the McAdam brothers established themselves across the country, their father began to consider his own finances. Until he was appointed surveyor to the Bristol Trust, all his work on the roads had been at his own expense. No one had paid the cost of his travel, let alone a fee, and after his system became widely adopted, he considered that the nation owed him a debt, especially the Post Office who had taken full advantage of his services. In 1819, therefore, he began a determined campaign to get himself a grant from public funds. The proceedings which followed, leading up to and beyond an enquiry by a select committee of the House of Commons in 1823, made McAdam, his family and his system, a centre of public attention, both friendly and hostile.

REFERENCES

Parliamentary Papers are :
Report from the Select Committee on Mr McAdam's Petition, relative to Road Making – 1823 (476) 5 53, cited as SC 1823.
Mr McAdam – Return of Salaries, Gratuities, or Remunerations received by Messieurs McAdam 1818–1825 – 1825 (248) Turnpike Trusts 1814–1833, Reports & Returns Vol 10 pt 2, cited as RR 1825.
Report from the Select Committee on Turnpike Trusts 1839 (295) ix 369, cited as SC 1839.

1. SC 1823 49.
2. Family information from Spiro; *Gentleman's Magazine* April 1853 (death of James McAdam Jr); papers belonging to the Rev H P Fox; Minutes of Exeter, Truro, Launceston Trusts (Christopher); Abstracts of Expenditure of Turnpike Trusts 1836 ff.
3. SC 1823 38.
4. As (1).
5. Mark xvi 15.
6. As (3).
7. Abstracts as (2); information from County Archivist, Cumbria; Spiro 417–448.

8. J L McAdam to the Duke of Richmond 13xi30, Goodwood MSS 1445/M40.

9. K. A. Macmahon, *Roads and Turnpike Trusts in East Yorkshire*, East Yorkshire Local History Society 1964, 55–6; Cambridge and Ely Trust Minutes 4x22, Cambs RO; SC 1823 Appendix Q.

10. Lemsford Mill and Welwyn Trust Minutes 19i20, Hatfield House.

11. SC 1825 4.

12. Ronald Good, *The Old Roads of Dorset* 1966, 145; RR 1825 5–6.

13. Truro Turnpike Order Book 1823–35, Minutes and correspondence April–June 1825, Cornwall RO.

14. Launceston Trust Minutes 7vi 19vii 1825; 29x 26xi 1829, Cornwall RO.

15. Cheshunt Trust Minutes, various dates. Herts RO.

16. N W Webster, *The Great North Road* 56–7. The Old North Road joined the Great North Road at Alconbury, approximately where A14 joins A1.

17. Kneesworth & Caxton Trust Minutes, South District, Cambs RO. James calls this trust 'Royston' in his list of trusts printed in SC1823.

18. Wadesmill Trust Minutes, Herts RO.

19. St Albans (Pondyards and Barnet) Trust Minutes, Herts RO.

20. RR 1825 3–4.

21. Estcourt's seat, New Park, is listed in Cary's *New Itinerary* 1828, 714; see DNB art. Estcourt, Thomas Henry Sutton; Hansard NS XIII 599.

22. SC 1823 26.

23. SC 1823 33.

24. SC 1839 Q413 34.

25. Truro Minutes 6x28; Launceston Minutes 6ix32.

26. Truro Minutes 13ii33.

27. AS (17) 2iv29.

28. SC 1823 46; James McAdam to the Duke of Richmond 10ii34, Goodwood MSS 1485/138, West Sussex RO; coaching information from London Stage Coach Directory in Cary, 1828, and *British Almanac* 1832.

29. SC 1823 39 46 49.

30. RR 1825 5–6.

31. Truro Minutes 5v28.

32. SC 1823 41.

33. RR 1825 2; SC 1823 49.

34. SC 1823 41; RR 1825 6.

35. SC 1823 41 46 55; Minutes of the Commissioners of Metropolis Roads 24x26, Greater London RO.

36. Lewis, *The Navy in Transition*, 87, 219.

V

Asking for Money 1819–1825

IN MARCH 1819 J L McAdam spent two and a half days before a Select Committee of the House of Commons, one of the long series which in the 18th and early 19th centuries enquired into turnpike road affairs. James McAdam gave evidence too, and so did Thomas Telford and three others concerned, as surveyor or engineer, with the upkeep of roads. Trustees from Cheshunt and Epsom, at both of which James was employed, gave evidence, and a land agent and civil engineer who said he was 'occasionally employed to solicit bills in parliament as an agent'.[1] From the side of the road-users evidence was heard from Charles Johnson, superintendent of mailcoaches, from three coach proprietors in London and from two on the Bath road, all being carefully distinguished by the title Mr from Charles Johnson Esquire.

J L McAdam was questioned at length about his methods, giving him an excellent opportunity, which he did not neglect, to display their merits. James was given a similar, though rather shorter, opportunity. Most of the other witnesses were either invited to comment on the value of the McAdams' work or did so of their own accord, and they were unanimously favourable. Whether this chorus of praise was orchestrated it is impossible to say, but it looks rather as though it may have been. Telford, the McAdams' great rival, did not join in, nor did the other witnesses professionally concerned with roads. On the other hand they offered no direct criticism, contenting themselves with describing their own methods which were different.

The Committee's report followed the tenor of the evidence :

'The general testimony borne to his [J L McAdam's] complete success wherever he has been employed and the proof that his improvements have been attended with an actual reduction of expense, while they have afforded the most useful employment to the poor, induce your Committee to attach a high degree of importance to that which he has already accomplished.'

They dismissed McAdam's somewhat muted plea for widespread consolidation of trusts and some kind of central inspectorate to oversee turnpike administration, but some of his views they heartily endorsed, particularly his proposal 'that each county or large district in the country ought to have an officer in the character of a gentleman, to oversee the surveyors of the district' – in other words, a county surveyor – and 'that no palliative, no other means whatever can be devised to get the London roads improved, except consolidating the trusts under one head'. They considered that 'high praise is due to the superior science exhibited by Mr Telford, in tracing and forming the new roads in North Wales', but in general the whole Report represented a triumph for McAdam, his sons and his system.[2]

McAdam by this time considered that he needed something more solid than public acclamation, welcome though that might be. He had spent a great deal of his private fortune on his entirely unofficial inspection and improvement of turnpike roads and so far he had seen no return except his salary as surveyor at

Bristol. Since his work was acknowledged to have been for the public benefit he put in a claim on the public funds, in November 1819, 'for payment of his expenses, and remuneration for his services'.[3]

Such a claim was by no means unprecedented. Sir Thomas Baring said in Parliament in 1825 that there had been upwards of twenty similar grants for similar public benefits in the previous twenty years.[4] Moreover, when it arrived at the Post Office from the Treasury, it was sympathetically received. 'As Superintendent of Mailcoaches,' commented Charles Johnson, 'I have abundant reason to wish that Mr McAdam's principles were acted upon very generally : if they were, a pace which in winter, or any bad weather, cannot be accomplished without difficulty, would become perfectly easy; to say nothing of the comfort and safety of the traveller, and the credit to humanity in lessening the labour of the animals.' As a postscript he mentioned 'as one instance of the benefit of Mr McAdam's improvement . . . that the *mail last winter lost ten, fifteen, and twenty minutes, in passing from Staines to Bagshot*, but now *the time is exactly kept*'.[5]

The Postmasters-General sent Johnson's remarks to the Treasury, adding some complimentary observations of their own on stretches of road which McAdam had improved. Their letter, dated 20th December 1819, was followed in February 1820 by a 'Certificate and Recommendation by several Peers and Members of Parliament, to the Right Honourable the Lords Commissioners of the Treasury, respecting Mr McAdam's Claim for remuneration'. The twenty-six signatories – sixteen commoners and ten peers, including both Postmasters-General, Earl De La Warr and the Earl of Hardwicke – made an impressive array.

The strength of McAdam's support and the urgency of his need induced the Lords of the Treasury to take the remarkable step of authorizing a grant from the Post Office revenues without waiting for Parliamentary sanction :

'. . . these testimonials', they wrote to the Postmasters-General on 22nd February, 1820, 'are of so highly respectable a nature from the station and character of the individuals who have signed them, and are decisive as to the merit, not only of the system itself, but also of Mr McAdam's personal labours and exertions in reducing it into practice . . . that my lords could not hesitate a moment in affording to any application which Mr McAdam may be advised to make to Parliament for remuneration of those services, their perfect and entire concurrence. But it having been represented to their lordships, that the pressure upon Mr McAdam . . . is of the most urgent description . . . my lords would be disposed to relieve Mr McAdam from his existing embarrassments, by the advance of any reasonable sum of money . . . leaving to the wisdom and justice of parliament, ultimately and finally to decide upon the extent of Mr McAdam's remuneration.'

The sum they suggested the Post Office might advance, from its own revenues, was £2,000.

£2,000 was welcome, but not nearly as much as McAdam thought he deserved merely for repayment of expenses, let alone 'remuneration', so he and his friends kept up their pressure. He had kept memoranda of his journeys, and when a Select Committee sat in 1820 to consider the state of the highways he took the opportunity to present a sworn statement of his travelling and expenses from 1798 to 1814. He had spent, he said, five years and 95 days in travelling 30,000 miles, as follows :

1798	1,115	1802	1,212	1806	634
1799	604	1803	300	1807	1,731
1800	800	1804	2,552	1808	2,382
1801	1,059	1805	1,641	1809	1,410
1810 1811 1812 1813	2,174	1814	1,350+		

For this travelling, 'reckoning by the rules of allowances made by the post-office to their surveyors', which he took to be £2 2s a day plus 2s (10p) a mile, he claimed £5,019 6s. Then, apparently as an afterthought, he put in a claim for £2,742 for his journeys after 1814. No record of the time and distance seems to have survived.

There are oddities about McAdam's major claim. Firstly, the mileage figures show not 30,000 miles but 18,964. Secondly, using 30,000 miles as a basis for calculation, the true total of the claim, at two guineas a day plus 2s a mile, would be £7,032 : using 18,964, it would be £5,928 8s. The arithmetic of the claim is thus obscure. Another oddity follows : both mileage and the claim based on it seem to have been accepted by the Committee of 1820, and yet they did not recommend that £5,019 6s should be paid. Instead, they recorded their opinion 'that Mr McAdam is entitled to further reward for his services, but they think it much better in all respects to leave the amount to the Post Office, than to mention any specific sum themselves.'

At the Post Office they were clearly sceptical of McAdam's major claim but unwilling, presumably because it had been favoured by the Select Committee, to press scepticism as far as outright challenge. 'We presume,' the Postmasters-General (Chichester and Salisbury) wrote, in a letter to the Lords of the Treasury, 'that your Lordships are satisfied that the claims of £5,019 6s "proved to the Committee" as the amount of his expenses to August 1814, are such as ought to be admitted for the ability, understanding and zeal which he has shown in his successful pursuit of the best means for constructing roads.'

So much for the larger claim. Passing to the smaller one, for £2,742, the Post Office authorities drastically scaled it down, incidentally destroying as they did so the basis of the larger claim, for they said that the travelling allowance for the surveyor and superintendent of mail coaches was £1 6s a day, not £2 2s. Taking the lower rate and adding 2s a mile for distance covered (they did not specify how much), they calculated the true sum due as £2,002 6s 'subject to the further reduction of £364 18s 6d being the sum he has received from various trustees, for expenses upon journies, included in the above account.' It was their duty, the Postmasters-General continued, 'to point out this difference in the rate of charges, amounting to £539 4s, leaving it to your Lordships to decide at what sum Mr McAdam ought to be paid, and confining ourselves to the expression of our opinion, that the allowances which have been made by the Postmasters-General to the surveyor and superintendent of mail coaches, for many years past are fair and reasonable.' This looks like a criticism, thinly veiled, of excessive claims by McAdam and of MPs' gullibility in accepting them.

'With great diffidence', the Postmasters-General nevertheless offered the opinion 'that if the further sum of £2,000 or £2,500 be allowed to Mr McAdam, in remuneration for those general advantages resulting from his plan, it will be a moderate compensation for [his] expense of labour and private fortune'. They added that from communication they had had with McAdam they had reason to believe that such a sum would be satisfactory to him: evidently he was prepared to strike a bargain. The Lords of the Treasury, in the autumn of 1820,* paid McAdam a second sum of £2,000, possibly intended as another instalment against his claim for expenses rather than the 'remuneration' suggested by the Post Office. However that may be, McAdam was still unsatisfied and he continued for more than two years to memorialize and petition for what he called 'the Balance due to him', and the Post Office continued patiently to support his case, though with diminishing enthusiasm.

Early in 1823 an application by McAdam reached the Chancellor of the Exchequer, Frederick Robinson (1782–1859). 'He was not disposed,' as he told the House of Commons in 1825, 'to receive it in the way which the friends of Mr McAdam could wish . . . He was . . . inclined to be rather stingy of the public money on this occasion,' and he refused the application; but he did consent to give the sanction of government to a petition of Mr McAdam to the House. Accordingly in the

* This is the date given of the issue of 'a second sum of £2,000' in the Report of the SC of 1823 (p. 7). Hansard (NS XII 1352 ff) says £2,000 was paid in 1820 and £2,000 'on the very strong recommendation of the Post Office' in 1822. There seems no way of reconciling the dates.

spring of 1823 a Select Committee under the banker, Sir Thomas Baring, sat to review the whole affair, being appointed specifically 'to take into consideration the Petition of Mr McAdam, and to report to the House whether any and what further pecuniary Grant shall be made to him, either by way of payment of his Expenses, or as a remuneration for his Services'.

The evidence given before this committee by McAdam and his sons has been freely drawn upon in previous chapters and need not be reviewed at length here. Broadly speaking, the Committee sought to establish how much and what kind of work McAdam had done on the roads, how much money he had paid from his own pocket in doing it, and how much he had received in fees, salaries and expenses. The committee sought similar information from all three sons and enquired whether the activities of father and sons, taken together, could in any way be considered a family business carried on for their joint account. McAdam strongly denied it. 'I conceive', he said, 'that the money my sons have is their own. I never even enquire about it, nor ever knew, till I came before this Committee, what my sons had. I did not even know the names of the Trusts that they have in their charge, so far did I conceive myself not entitled to interfere in their private affairs.'[6]

This was disingenuous, to say the least. It was true that McAdam & Sons were not formally in partnership with each other, but how could J L McAdam say that he did not know which trusts his sons were working for when he had frequently been instrumental in placing them in employment? Moreover, in correspondence he and William frequently used the word 'we' in contexts which suggested a family business, even if not a formally constituted one.

Apart from the McAdams, the Committee heard several witnesses in their favour and on the other side two professional rivals: William Lester who claimed in a rather muddled way to have anticipated McAdam's methods and to have invented various machines for road-making, and Benjamin Wingrove, surveyor to the Bath Trust, of whom more will be said in Chapter VI. Wingrove had quite a long hearing but his claim for compensation for his services was dismissed as 'foreign to the subject of the enquiry', and Lester's also.

The tone of the Committee's report, dated 20th June 1823, is throughout strongly favourable to McAdam. On the main terms of reference it says:

'the Committee cannot hesitate to express their opinion, in concurrence with that already pronounced by the Heads of the Department of the Post Office, that the sum of £2,000 or £2,500 in addition to his [J L McAdam's] expenses . . . will be but a moderate compensation to Mr McAdam for his great exertions and very valuable services.'[7]

It was nearly two years before this recommendation came up for debate, on 13th May 1825. The Chancellor, by then, had changed his mind, partly as a result of talking to members of the Select Committee, partly as a result of the report itself, and he consented to propose a grant of £2,000. 'It might be said,' he observed, 'that when any man made an ingenious discovery he would be rewarded by the employment that must necessarily follow. But that was not always the case . . . In the present instance . . . he really thought Mr M'Adam was entitled to further remuneration; because, although not in a state of poverty or destitution, the advantages which he had derived from his exertions were by no means adequate to their importance.'

The grant was far from universally popular. The future Lord Chancellor Brougham (1778–1868), speaking a month earlier on a successful motion to have the debate postponed, said that if the motion for the grant were carried 'there was no one invention made . . . which might appear as beneficial to the country, that might not be advanced as a good ground for a grant of public money'. McAdam's originality was challenged by Joseph Hume (1777–1855) the Scottish radical, who supported Lester's claims, and by H Sumner, one of the members for Surrey. Sumner also said that the McAdams had already drawn £41,000 from different trusts over the previous five years (he did not explain his calculations) and that their

claims for expenses were 'made, not on the economical rate of a
surveyor of the road, who would have been satisfied to ride
his horse and dine on a beefsteak and mutton chop. It would
seem that the firm of M'Adam travelled in their post-chaise and
four, and enjoyed all the delicacies of the season'. Sir M
Cholmeley thought the £4,000 that McAdam had already
received was enough, Sir E Knatchbull feared that after this
grant 'they would, ere long, be applied to for further remuneration to Mr M'Adam'. The member for Devizes, as we
have seen (see p. 94) attacked William McAdam's management
of the Devizes roads.[8] Hansard's report suggests that the
opposers of the motion were more eloquent than its supporters.
Nevertheless, when it was put to the vote in a thinnish house it
was carried by fifty-six votes (eighty-three to twenty-seven),
and J L McAdam was assured of his third grant of £2,000.

So far as it is possible to tell, McAdam's total income from
twenty-five years' work on the roads (not all of it full-time)
was the £6,000 of public money which he received between
1820 and 1825, his salaries from his appointments at Bristol,
Bath, and in the North, fees and expense allowances from a few
other trusts, and perhaps something from his writings. These
receipts, he calculated, were too small to cover his expenses,
so that his labours, far from enriching him, had left him out of
pocket. His widow, in 1840, quoted by Spiro, told the 10th
Earl of Dundonald that he was unable to leave her 'more than
the mere amount of salaries due at his death, which were but
trifling'.

Some of his contemporaries found this hard to believe, and
in an age of family businesses they found it even harder to
credit the notion that the earnings of McAdam's sons were
separate from his own. On this point, as on others, it has been
shown that McAdam was not an entirely reliable witness. No
one, however, challenged him during the parliamentary hear-
ings, and those who accepted his own version of his affairs –
senior officials, politicians, members of the unreformed House
of Commons – are unlikely to have been unduly credulous, so
that it may be reasonable to conclude, with them, that what

McAdam said was substantially true. They were unwilling, however, to grant him a sum anywhere near the amount of his full claim, which suggests that although they may have given him the benefit of the doubt, a doubt still remained.

By the time the Committee on McAdam's Petition reported, McAdam's fame was rising to its height and spreading very widely. He was being noticed abroad, and in 1823 the first American road constructed on his principles was opened between Hagerstown and Boonsboro, Maryland.[9] Between 1824 and 1826 the noun *macadam* and the verb to *macadamize* came into the language.[10] As his reputation grew, criticism of his methods, which may appropriately be considered along with the question of his reward, began to emerge, associated with the name of his close contemporary, the greatest road-builder of the day, Thomas Telford (1757–1834). It was directed not at his principles of administration, which were unassailable, but his directions for forming a road surface – at 'macadamization' itself.

In evidence to the Select Committee of 1819,[11] Telford was as scathing as McAdam about traditional methods of road building and maintenance, which in his opinion showed a total lack of skill or proper care in drainage and in the use of materials. He recommended, much in the same way as McAdam, that stones laid on the road should never weigh more than six or eight ounces each; that the surface should have 'a very gentle curve in the cross section'; that if a road were built on 'clay, or other elastic substance, which would retain water', it should have a layer of 'vegetable soil' between the surface and the clay. Gravel, he said, should be thoroughly cleaned and round stones should be broken with 'a small hammer'. On this last point, which McAdam also stressed, he was most insistent, saying that the 'weight, shape and manner of using' hammers was 'of much more importance than any one can conceive who has not had much experience in road-making'.

Telford's disagreement with McAdam, which does not emerge in the 1819 evidence, was over the matter of foundations beneath the surface of the road. The small stones of the metalling, he considered, needed 'a rough pavement' under them, 'or otherwise, a portion of the broken stone metalling comes in contact with the earth, sinks into it, works unequally, and can never be rendered so perfectly uniform as when the layer of broken stone is placed upon a proper pavement'.[12] No one spoke with greater authority than Telford, for his experience of road-building in Scotland, in England, and in the mountains of Wales extended over a great many years and about as great a mileage as that covered by the entire McAdam family.

J L McAdam, nevertheless, put forward the startling theory, which he maintained almost to the point of perversity, that no foundation was needed at all under the metalling: the metalling, indeed, might be better without it. Let us listen to him for a moment, answering questions put by members of the 1819 Committee, who were obviously incredulous:

Q: What depth of solid materials would you think it right to put upon a road, in order to repair it properly?
 – I should think that ten inches of well-consolidated materials is equal to carry anything.
Q: That is, provided the substratum is sound?
 – No; I should not care whether the substratum was soft or hard; I should rather prefer a soft one to a hard one.
Q: You don't mean you would prefer a bog?
 – If it was not such a bog as would not allow a man to walk over, I should prefer it.

Faced with this astonishing pronouncement, a member of the Committee asked McAdam what advantage was derived from the substrata not being perfectly solid, and McAdam said he thought that a road built on a hard substance 'such as rock' wore much sooner than one on a soft surface, and that the difference in drawing carriages over the two roads would be very slight. This point the Committee pursued, to get, perhaps, the most surprising reply in the whole dialogue:

Q: To use the expression to which you have alluded, as being used by the coachmen, would a carriage run so true upon a road, the foundation of which was soft, as upon one of which the foundation was hard?

– If the road be very good, and very well made, it will be so solid, and so hard, as to make no difference. And I will give the Committee a strong instance of that, in the knowledge of many gentlemen here. The road in Somersetshire, between Bridgewater and Cross, is mostly over a morass, which is so extremely soft, that when you ride in a carriage along the road, you see the water tremble in the ditches on each side; and after there has been a slight frost, the vibration of the water from the carriage on the road will be so great as to break the young ice . . .'

McAdam went on to claim that the upkeep of the 'soft' road cost only five-sevenths of the upkeep of the 'hard' road, and he repeated that coachmen found no difference in running upon the one and upon the other. In forming a road over a morass, he was asked, would he bottom it with small or large stones? 'I never use large stones on the bottom of a road,' he replied, 'I would not put a large stone in any part of it' – nor would he use faggots or any other kind of intermediate material between the surface and the road-bed. Worn out, perhaps, by collision with Scottish granite, the Committee eventually veered away towards less treacherous ground: the proper size of stones for surface metalling.[13]

McAdam's line of argument laid him open to attack from those who maintained that his whole approach to road-making was unscientific – as, strictly speaking, it was, being based on generalizations drawn from intelligent observation rather than from controlled experiment. Sir Henry Parnell, his most formidable opponent at the governmental level, maintained that his unsupported roads with their few inches of metalling were unduly weak for the increasing traffic of the day and that an 'elastic' surface such as McAdam advocated must, on scientific grounds, be more difficult for a vehicle to run upon than a hard one.[14] As the caption to an exhibit in the Smithsonian succinctly puts it: 'A horse can draw more than twice as much on macadam, and more than three times as much on good Telford-Macadam' – that is, presumably, on a road with a macadam surface and Telford's paved foundation – 'than . . . on a gravel road'.

McAdam would have done better to stand on the ground of economy, always a cardinal consideration with turnpike trustees. Telford's method no doubt gave a stronger road, but a road much more expensive to construct. Nor, in practice, did McAdam's roads turn out so weak as Sir Henry Parnell suggested, nor so hard to run upon. On the contrary, they carried the traffic of the day – heavy and increasing – perfectly adequately, and their surface could be put down and kept in order for an outlay well within the means of a properly managed trust. McAdam, as he frequently pointed out, would provide the management as well as the surface.

McAdam's system of management had, nevertheless, one weakness. He and his sons pointed out time and again, in reports to turnpike trusts and elsewhere, that roads maintained in the traditional manner had so much unbroken stone dumped on them that if McAdam's system were adopted it would be unnecessary for years to quarry or cart any fresh stone, and so the cost of upkeep would be bound to drop. It would no doubt rise again when the original stock of materials ran down, but that, unless pressed, the McAdams forbore to point out, leaving their rivals to do it for them. 'The manner . . . in which he [McAdam] produces the saving of expence', said Joseph Mitchell, one of Telford's assistants who from time to time came into conflict with McAdam in Perthshire, 'is in many cases utterly delusive . . . On acquiring the charge of a Turnpike Road . . ., he saves the expence of quarrying and carting new materials by breaking up the surface of the Road . . . and relaying it in a regular and uniform shape. In this manner he makes a better Road and creates a striking saving of expence – This continues until the whole stock of materials is exhausted, but by that time he is fairly established in the Trust, and can come forward with a tolerably good grace for additional supplies.'[15]

The Royal Mail passing Temple Bar, 1834.

Museum of London

'English direct and cross mail' in the time of William IV (1830–1837),
being transferred between the coach and a mail cart for local delivery.

Fotomas Index

Stage Coach: Opposition Coach can be seen coming into sight on the
left of the picture.

Fotomas Index

Making due allowance for professional jealousy, the substance of this criticism seems fair. As material already on the
roads was used up, there would be a tendency for upkeep costs
to rise : not a matter which the McAdams were at pains to raise
when they made their bid for a trust's business. The Bath
trustees had the happy experience of finding in 1829, after J L
McAdam and his grandson had been in office nearly four years,
that they could reduce their annual allowance for repairs from
£7,500 to £6,900, and they later presented their surveyors
with 'their strongest (and unanimous) thanks for the zeal,
efficiency and integrity, with which they have . . . so amply redeemed every expectation', but by no means all trustees were so
satisfied.[16] Some went back to contractors and the St Albans
trustees, who seem always to have been determined to keep
James in his place, perhaps because they had been amongst his
earliest patrons, accused him in 1832 of not having carried out
his original estimate, made in 1823, that one-third of the
expense of management would be saved, and he did not directly
refute the charge.[17] Mischances of this nature were no doubt
what prompted Mitchell's caustic observation : 'Trustees of
Turnpike Roads . . . are carried away by the appearance of
striking economy at the outset, and have employed Mr McAdam
and his family in every direction.'

It is also true, no doubt, that McAdam's roads were less
robust than Telford's. On the other hand they cost less to
construct, and on the general issue of economy and efficiency
even Mitchell was forced to admit that 'no small credit is due
to them [the McAdams] for the superior order and regularity
which they have introduced into the expenditure of Trust
Funds, and into the detail of Road Repair generally'. For the
£6,000 which McAdam eventually extracted from the Treasury
the nation probably got as good a bargain as it has ever had
from a grant of the taxpayers' money.

Macadam

REFERENCES

SC 1819 – Report from the Select Committee on the Highways of the Kingdom – 1819 (509) 5 339.

SC 1823 – Report from the Select Committee on Mr McAdam's Petition, relative to Road Making – 1823 (476) 5 53.

1. SC 1819 52.
2. As (1) 4 and generally; evidence of J L McAdam.
3. SC 1823 3.
4. Hansard, *Parliamentary Debates*, New Series XIII, 597.
5. For this and succeeding paragraphs see SC 1823, Report 3. *et seq*, and Appendices, especially Appendix H for Postmaster-Generals' letter to the Lords of the Treasury, 18x1820.
6. As (3) 55.
7. As (3) 8.
8. For Brougham, Hansard NS XII 1352; remainder, as (4) 593 *et seq*.
9. Smithsonian, Washington DC.
10. Shorter OED.
11. As (1), Telford's evidence, 54–7.
12. Telford to William Adam June 1832, quoted Gibb, *Story of Telford* 176.
13. As (1) J L McAdam's evidence 23–4.
14. Sir H Parnell, *Treatise on Roads*, 24–5, 64–5, 76–7.
15. As (12) 177–8.
16. Bath Trust Minutes 7x29 4vi31.
17. St Albans Minutes 28iii32, James's reply 27iv32.

VI

Macadamizing the West of England

The Bath Campaign 1817–1826

IN THE 1820s and early 1830s long-distance road transport was briefly at the height of its horse-drawn speed and efficiency before the railways destroyed it, and at the same time McAdam and his methods were at the height of their prestige. During these years two wealthy and active West of England trusts, at Bath and Exeter, employed four McAdams, of three generations, as surveyors. Their plentiful surviving records illustrate the work of John Loudon McAdam, of his son William, and of William's two sons William Jr and Christopher – John Loudon and William Jr at Bath : at Exeter William Sr and Christopher.[1]

The Bath trustees had charge of about 49 miles of hilly roads (see map on p. 120) radiating from the city westward towards Bristol and South Wales, northward towards Gloucestershire and the Midlands, eastward towards London, southward through Somerset coalfields towards Wells and the South-West of England. The traffic on these roads, said the trustees in 1825, 'from their Contiguity to the two great Cities of Bath and

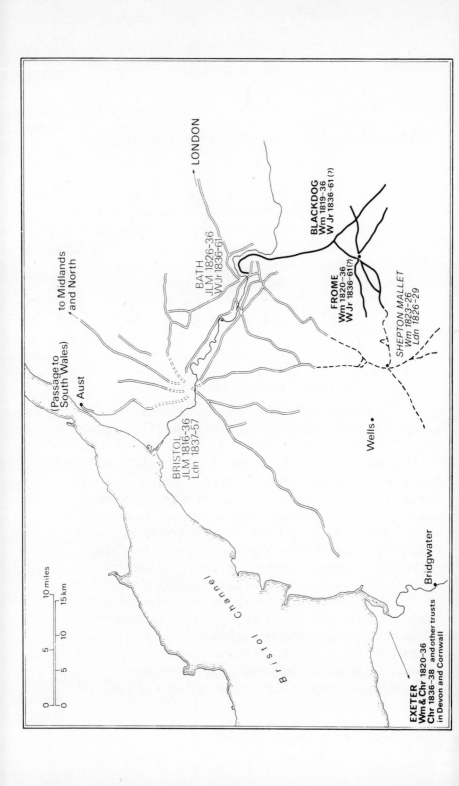

LONDON →

BLACKDOG
Wm 1819–36
W Jr 1836–61(?)

BATH
JLM 1826–36
W Jr 1836–61

to Midlands
and North

FROME
Wm 1820–36
W Jr 1836–61(?)

SHEPTON MALLET
Wm 1823–26
Ldn 1826–29

(Passage to
South Wales)

● Aust

BRISTOL
JLM 1816–36
Ldn 1837–57

Wells ●

Bristol Channel

Bridgwater ●

EXETER
Wm & Chr 1820–36
Chr 1836–38   and other trusts
in Devon and Cornwall

10 miles
15 km

0   5   10

0   5   10   15

Bristol, and the Population of the Districts through which they extend, is unusually great, [and] the Bath Trust is peculiarly affected by two Articles of Consumption, which are commonly supplied to other populous Places by Means of Navigations or Rail-Roads,* viz. *Coals* and *Building Stone*; the neat Quantities of which, exclusive of the Weights of the Carriages employed, are calculated at about Eight Hundred Tons per Day, and pass over, in different Lines, nearly thirty Miles of Road, many of which consist of long Hills, notwithstanding a Canal has long since been made and formed for supplying the City of Bath with Coals by Water Carriage.'[2]

Besides this heavy goods traffic, coaches ran all day between Bath and Bristol (see p. 43). Between Bath and other provincial towns, in 1828, there were thirty-one coach services, of which eight ran daily and twelve ran daily excepting Sundays. Fourteen services linked Bath with London, including the overnight Exeter mail and another night coach serving Bristol as well as Bath. The fastest time from London was twelve hours, with two hours on to Bristol, giving an average speed, including stops, between eight and nine miles an hour.[3]

During the years between 1810 and 1816, when J L McAdam's influence was rising towards its zenith, the Bath Trust was prosperous and its numerous trustees were active and quarrelsome, thereby largely nullifying any good results that might have flowed from their undoubted energy. As early as 1810 one of them, the Rev. Mr James of Radstock, was anticipating McAdam's methods by having the roadstone near him 'broken to the Size of a large Walnut', much smaller than anywhere else in the Trust's territory. There was much animated discussion of the bad state of the roads and of measures needed to improve them, particularly the appointment of a General Surveyor or Surveyors, but so much ingenuity was displayed in the manoeuvres of various factions to block each other's plans that no appointment was actually made until October 1815. By that time Benjamin Wingrove, the son of a Bath brewer and coal merchant and an active and bellicose trustee,

* Presumably worked by horses.

had thickened the brew of venom by publishing an attack on 'the laches and erroneous proceedings of the Turnpike Board, or of those, or *some of those*, to whom that Board has delegated its authority'. It was in such an atmosphere, late in 1816, that the trustees, probably partly at Wingrove's instigation, decided to write to 'Mr McAdam the Surveyor General of the Bristol District of Roads requesting his attendance . . . to confer with the Trustees relative to the State and Improvement of the Roads under this Trust.' 'In the management of roads', Wingrove had said in an address published in 1815, 'they [at Bristol] are at least a century before us' – and that was while McAdam was a Bristol trustee but before his appointment as Surveyor.[4]

On 21st December 1816 McAdam met the Bath trustees. He 'stated the general outline of Improvement in the repair of Turnpike Roads in general and obligingly offered to make a Survey of the Bath Roads and ascertain the best means of repairing the same and managing the funds of the Trust in future'. In what appears to have been a general outburst of goodwill Wingrove was asked to help and thanked for 'his Exertions in bringing the Subject of the Hotwell Stone' – which had led to the invitation to McAdam – 'before the Board'.

Almost immediately McAdam set to work, supported by a team consisting of his son William, 'a surveyor and one or two workmen from Bristol and some Bath labourers. He installed himself at the Bath Inn from which, in a phaeton and pair, he and others 'travelled on the Bath district, several times over, the 49 miles of road, examined all their quarries, pitted the whole of the road to ascertain its strength and condition, and were obliged, from our own affairs at Bristol, to go several times backwards and forwards between Bath and Bristol'.[5]

He stayed at Bath until 15th January 1817 and his report was ready for a trustees' meeting on 1st February. It survives, and it is a standard McAdam document. The results of pitting at 303 'stations' were set out in detail and from them, McAdam said, 'it appears that the whole of the Roads are sufficiently strong, and that most of them have been overloaded with

materials'. The system of repair was defective 'both as respecting the form of the roads, and the preparation of materials' and
the watercourses were 'unjudiciously conducted'. The finances
of the trust would allow £162 2s 10d a week (£8,327 7s 4d a
year) for upkeep and that, McAdam thought, would be ample
once the roads had been brought to 'a tolerable state of repair',
but 'the application of a sum of ready money' – £1,500 – would
be needed first. Once the roads were in 'a smooth solid state,
and . . . a shape the most convenient to be travelled upon', it
ought to be possible to keep them in that condition for £5,500
a year.

Characteristically, McAdam's remarks on the physical state
of the roads were followed by recommendations for a system of
management. He deplored the trust's prevailing practice of
leaving considerable stretches in the care of the parishes and
proposed, instead, the appointment of a General Surveyor and
three sub-surveyors who, at 1½ gns (£1.57½) a week each,
would each have charge of a district. It would be more expensive
than paying a total of 63s a week for five sub-surveyors, but
'these men neither have, nor pretend to have, any skill or
judgment to direct the workmen; they are considered merely as
overseers to prevent idleness', and McAdam was confident that
his system was the right one. As to the General Surveyor, he
burst into eloquent peroratory:

> 'The success of the great object of the Commissioners, in
> the amelioration of the roads under their care; the economical
> management of their finances; and the consequent relief and
> advantage of the commerce and agriculture of the city of
> Bath, and the surrounding country, must ultimately depend
> on the ability, zeal, and firmness of the officer to whom they
> shall commit the execution.'[6]

McAdam was talking about himself – who else? He expected
to add Bath to Bristol and he was encouraged all the more
when, at a second meeting to consider his report, his only likely
rival – Wingrove – withdrew a proposal of his own to avoid
clashing with McAdam and was thanked 'for the handsome

manner in which he has withdrawn the same'. The trustees then resolved 'that the principles as stated in such [a] report, and Mr McAdam's System of Road repairing be adopted and carried into practice on the several roads under this Trust'. They went further, and resolved to give 'the present Surveyors of the Roads' notice of dismissal on 25th March, just over five weeks from the day of their meeting, 15th February 1817. They asked McAdam whether he was ready to take charge of the Bath roads as general surveyor, and he said he was, adding that he proposed to place one of his sons permanently in Bath 'from a sentiment of respect to the commissioners, and as an eligible means of communicating with them at all times individually'.[7] This question and answer are not recorded in the trustees' minutes, but McAdam evidently took them to mean that the appointment would be his, without opposition.

That assumption was premature. Immediately after the meeting, said McAdam a month later, he learnt that 'another gentleman had presented himself as a candidate for the office, and that he grounded his pretensions on the impossibility of my fulfilling your wishes' – that is to say, that McAdam would not be able to serve Bath as well as Bristol. 'Another gentleman' was Wingrove, revealing himself suddenly not as a supporter of McAdam but as his rival, and during the four weeks following the February meeting the City of Bath and its neighbourhood must have been buzzing with local politics, for when the next trustees' meeting fell due, on 15th March, 109 trustees arrived at Bath Guildhall to settle a contested election to the office of General Surveyor.[8]

McAdam was deeply disturbed by Wingrove's intervention, feeling that his honour, of which he was always jealous, had been impugned. On the day when the vote was to be taken he submitted to the Bath trustees a letter, already referred to earlier, in which he made it quite plain that he considered that he could do the job, notwithstanding his responsibilities at Bristol, and that the trustees, having accepted his report, ought to appoint him. 'When I made that offer,' he said, referring to his reply to the trustees on 15th February when they asked if he was ready to take charge of the Bath roads, 'I had well considered my means of executing it with fidelity; and, to secure that object, I have called from Scotland my eldest son, now thirty-four years of age. With the assistance therefore of my family, I am enabled to undertake even a greater charge than the united Trusts of Bath and Bristol. Had there been any doubt respecting my ability to perform the service properly, I hope I have sufficient self-respect to have hesitated in making the offer.'

He went on to point out that he had given the trustees his opinion and plans 'in the most unreserved manner' and they had adopted them, so that he was identified with their future proceedings, as he put it, in a manner from which he could not extricate himself: failure was bound to throw discredit on him. 'I have no intention,' he said, 'of withdrawing myself from the responsibility of the measures I have recommended in my report; it is my sincere desire that I may be allowed to add the improvement of the Bath district to that of Bristol, by which a practical proof may be given to the country and to the legislature, of the necessity of giving to commissioners of turnpikes the assistance of an executive department.'

This letter was presumably distributed to the 109 trustees when they assembled at the Guildhall, and at first they seemed to go out of their way to show their confidence in McAdam, for they resolved to rescind and revoke all orders, resolutions and the like 'that may be in opposition to Mr McAdam's System'. Nevertheless, when the principal motion was put to the meeting, fifty-five trustees voted for Wingrove and fifty-three for McAdam, giving Wingrove a clear, if close, majority, for only one trustee abstained. Wingrove, accordingly, was appointed, with a salary of £350 including the expenses of an office in Bath.

The trustees, unabashed apparently by the tone or substance of McAdam's letter, issued instructions a few weeks after Wingrove was appointed that McAdam's system was to be adopted, and committees were set up for Inspecting the Execution thereof. Wingrove's own views, as he explained when he gave evidence to the Select Committee of 1823, were by now swinging against McAdam's.

When McAdam was first approached to report on the Bath roads, Wingrove's praise was fulsome – 'I only know him by his works, which far transcend in merit all that I have ever witnessed, though I have seen and inspected a very great proportion of the public roads in England and Wales.' Then McAdam, son, and entourage came to Bath to make their survey. Wingrove had been asked to help, but in spite of a note from him to say that he would be away from Bath for a few days McAdam went ahead and finished his work in Wingrove's absence: was that thoughtlessness, impatience, or a calculated affront? It may in any case have had some effect on what followed, seeing that Wingrove chose to recall it over six years later.[9]

After Wingrove had seen McAdam's report he entertained, he said, 'a very different opinion of the judgment of Mr McAdam as a road surveyor'. He claimed that he tried McAdam's system as the trustees directed, remarking in passing 'his system is not explained', but that it failed, and he proceeded to explain why.[10] His criticism had much in common with Telford's. 'I conceive no road formed upon such a bottom as is reported by Mr McAdam, could be of sufficient strength at six or seven or eight inches, for the usage to which these roads are subjected.' McAdam, he said, used poor materials, recommending freestone at eighteenpence a ton rather than Bristol stone at six or seven shillings, 'but the one will make a good road, and the best road, and the other I conceive it to be impossible to make a good road of, to last any time; I mean, because by the effect of the seasons alone, the freestone is utterly destoyed, even without the co-operation of friction'.[11]

Wingrove said that he sent for blue lias limestone 'from a considerable distance, and at a considerable expense' to repair hilly roads near Dunkerton which carried 'an immense quantity of coals . . . and the hills being steep, there is constant dragging'. He used also Hotwell rock-stone on 12 or 13 miles of road round the city of Bath. His system, he said, consisted 'first in the construction, which includes a good foundation; secondly, in the selection and preparation of the materials; and thirdly, in the system of drainage necessary for protection'. In its reliance on good foundations Wingrove's system resembled Telford's. It differed from McAdam's, he said, in giving the roads 'a little more convexity, but I have observed lately, that Mr McAdam's roads are not, by any means, upon the plan which I have seen recommended in some publications, under his signature'.[12]

Wingrove was active in propagating his view – he published quite frequently – and he was evidently a respected surveyor whose services, though less widely in demand than McAdam's, were nevertheless valued by several trusts in the West of England. In 1819 he was appointed surveyor to the Taunton Trust, a very large one with nearly a hundred miles of roads, after surveying them and recommending 'a very comprehensive system of management for the whole district'. His report was printed and attracted a good deal of attention, especially at Tiverton. A copy survives in the Somerset Record Office. W Kinglake, the clerk to the Taunton trustees, praised Wingrove's work highly to the Select Committee of 1823 – 'The trust can now boast of having, perhaps, the best roads in the West of England, and what is of greater consequence, the *science* of road-making is, through you, diffused through the district.' He failed, nevertheless, to get an appointment at Wells in 1819 in spite of pointing out that he was already in office at Bath and at Taunton and could thus provide a comprehensive service. He evidently had the ambition of building up a group of surveyor-ships in the McAdam manner and, on a much smaller scale, he was successful, becoming surveyor in his later years at Bradford-on-Avon, Chippenham and Trowbridge. He probably died in 1839.[13] His son Anthony was employed jointly with him at Bath in a partnership very reminiscent of the McAdams.

Competent though Wingrove was as a surveyor, he was by no means McAdam's equal in management. For some years his work for the Bath Trust went smoothly, but in March 1823 a well-attended meeting of trustees – thirty-four were present –

heard that while Benjamin Wingrove was away his son Anthony, by his direction, had signed bills payable to sub-surveyors in his name. The meeting decided that the elder Wingrove had been acting 'with a view to furthering the interests of the Trust', so they let the matter drop and elected Anthony joint Surveyor with his father, thus implicitly endorsing his proxy signatures.

The General Surveyor's accounts, however, were in confusion. They were investigated in detail, 'irregularities' were found, and the matter was publicly aired when Wingrove unwisely thrust himself before the Select Committee considering a grant to McAdam. The trustees, nevertheless, were charitably disposed, accepting Wingrove's explanation that the irregularities were due 'partly to the great labour of his office, partly to domestic affliction and principally to his trusting to others what he ought himself to have performed, but certainly not to any dishonourable or interested motive; and from the great improvements to our roads under his management, we do not feel inclined to say more than merely to express our regret'.[14] Twice during 1823 the trustees went further, expressly recording their approval of Wingrove's work, and on the first occasion, in June, they sent a copy of their resolution to the MPs for the county of Somerset and for the city of Bath. On the second occasion, in October, they were rather more measured in their approbation: *'For the most part'**, the trustees said, '[they] feel authorized to pronounce the Roads in excellent Condition; and they cannot close their Report without expressing their Approbation of the Exertions and Attention of their General Surveyors.'

In the light of later developments the italicized phrase looks ominous. Wingrove's enemies and McAdam's friends among the trustees, defeated in 1817, did not give up the fight, and Wingrove's incompetence with the accounts gave them a lever which they repeatedly used against him. At a statutory special meeting to deal with the accounts, in October 1825, when fifty-one trustees were present, twenty-eight voted for a

---

* Present author's italics.

resolution thanking the Wingroves for their work, fifteen opposed it, and John Allen gave notice of a petition to Parliament alleging mismanagement of the funds of the Trust. On 5th November the accounts were accepted and signed and no more was heard of Allen's petition, but the other side of Wingrove's job, care of the roads, came under attack. 'The larger Portion of the Roads is in good Repair,' say the Minutes, 'but there are other Parts not in good Order; and some few Parts in a bad state of Repair.'

Three weeks later Anthony Wingrove, speaking of a committee sitting to consider a report of the surveyors, was reported as saying 'that part of the Committee were Blackguards and part of them Gentlemen'. When he was taxed with the outburst he admitted it, but on Christmas Eve the trustees seasonably took account of the time which had elapsed and did not deem it necessary to call on him for any further explanation. The elder Wingrove at the time was ill, which could very well account for Anthony's anger at what he might have considered persecution of a sick man.

McAdam's friends, no doubt Anthony Wingrove's 'blackguards', were not put off by any feelings of tenderness for his father. On 14th January 1826 a Special Meeting of Trustees, with thirty-eight present and an Archdeacon in the Chair, decided that 'as a Measure of practical Utility and in order to put out of Doubt which of two opposite Systems can maintain... good and sound roads at the smallest Cost', one district should be maintained on McAdam's system. A committee was to be appointed to see that the trial was carried out, separate accounts were to be kept, a surveyor was to pit the roads at the start 'and also at the End of the Experiment in the presence of Mr B. Wingrove', and piles of stones were always to be measured in the presence of one of the General Surveyors.

Mr B Wingrove did not wait for 'the End of the Experiment'. A fortnight after the trial had been agreed upon, at a meeting attended by thirty-eight trustees, both Wingroves tendered their resignations. They were accepted, though not without some dissentient votes.

Macadam

The Chairman of the committee appointed to see to the trial of McAdam's system was Captain G T Scobell, RN. It was a courtesy captaincy. He became a Commander on 1st February 1812, appears never to have gone to sea in that rank, and was never promoted further: not necessarily a reflection on his professional abilities since naval promotion went by seniority and unemployment among Commanders, heavy even in wartime, stood at 92.1 per cent in 1824.[15] As a turnpike trustee he was assiduous, a redoubtable politician, and a persevering supporter of McAdam.

McAdam in March 1826 was in his seventieth year and a widower, his wife having died in February 1825. Whether there was much warmth between them may be doubted, for her name does not appear upon his tombstone. Perhaps he was more attracted by her much younger cousin Anne Charlotte De Lancey, who had a house at Hoddesdon in Hertfordshire, for in 1827 McAdam moved there from Bristol and in the same year married Miss De Lancey, then aged about forty. She was well-connected and in 1839 her brother W H De Lancey (1797–1865) became first Bishop of Western New York. McAdam's relations, nevertheless, were most upset by his second marriage, considering him too old for that sort of thing.[16]

In spite of this evidence of continued vigour, McAdam might reasonably have been considered too old for any new professional engagement before he left Bristol. That, however, was not the view taken in certain circles at Bath. As soon as the Wingroves resigned, the vacancy in their post was advertised, but at a meeting of fifty-nine trustees on 25th February 1826 all five applicants were rejected. Captain Scobell and the Revd Mr Seymour thereupon suggested asking McAdam to recommend a surveyor and to spend one week in each quarter himself superintending and directing the management of the roads and the state of the accounts.

There was considerable opposition, and at the next meeting of trustees on 11th March, when fifty-seven were present, attempts were made to get the approach to McAdam annulled as irregular and to appoint a man called George Layng Miller as surveyor. Both were defeated. Captain Scobell and Thomas Shaw sailed to victory with a decision to appoint J L McAdam General Surveyor from 25th March 1826 with a salary of £600 a year.

From the remaining terms of the motion it becomes apparent that although Scobell and his party were attracted by McAdam's prestige and were determined to have him, yet they did not expect him to spend more than part of his time on the affairs of the Bath Trust. His salary was set high so that he could afford to pay a deputy out of it, and he was engaged 'to be responsible for the Duties of the Office and himself or a competent Deputy to be usually superintending the Roads and Accounts'. On these terms McAdam united the surveyorship at Bath to his surveyorship at Bristol, undoubtedly a long-standing ambition. Financial constraints were imposed. McAdam undertook to keep the Bath roads in order for no more than £7,500 a year, including his own salary and the salaries of sub-surveyors. On these terms he was appointed for five years 'in order that the Trust may be benefited by this reduced expenditure during the whole of such period'. And, the minutes might have added, there would be plenty of trustees, hostile to Scobell and his nominees, waiting to pounce if the Trust did not get the benefits contracted for.

The 'competent deputy' mentioned, but not named, in the minute of 11th March was J L McAdam's twenty-three-year-old grandson, William, who remained with the trust certainly until 1860 and probably until he died in the following year. The elder William, whose tenure of surveyorships under some trusts was short and stormy, installed himself firmly from 1819 at the Black Dog Trust, which had an important road running into Bath from Warminster, and by 1834 his son had joined him as deputy, remaining as General Surveyor after William Sr's death until 1851 or later.[17] Black Dog connected with Frome, where William Sr was surveyor in 1820, and Frome with Shepton Mallet, where he became surveyor in 1823, not altogether to the trustees' satisfaction, and was succeeded by his brother Loudon before 1826. The Shepton Mallet roads were also once inspected by Loudon's nephew James Loudon John, son of his elder sister Ann, and her husband Captain James Saunders, RN. Neither seems to have pleased the trustees much better than William, and Loudon was dismissed in 1828 after interrogation of his sub-surveyor, recorded verbatim in the minutes, revealed that he had only visited the trust three times in three years and his nephew once, and that for twelve months neither of them had been seen.[18] Probably the McAdams were attempting too much, neglecting the smaller trusts for the greater, though Shepton had 51 miles of road and about £3,500 income. In the mid-1820s the influence of the family ran unbroken over a network of about 340 miles of road in South Gloucestershire, North Somerset and West Wiltshire. Shepton was soon lost, but after that it seems certain that the remaining 289 miles were held by McAdam hands as long as there were McAdam hands to hold them.

William McAdam & Sons in Devon and Cornwall,
1818–1846:

Further West, between 1820 and 1837, there was another large province of the McAdam empire. Before passing to the work of J L McAdam and his grandson at Bath it will be convenient to consider the rise and decline of William and his son Christopher as surveyors to the Exeter Trust.

The Exeter trustees managed about 150 miles of road – about three times as much as the trustees at Bath – radiating in all directions, and J L McAdam said that Exeter had the largest trust in the country after Bristol.[19] Against Bristol, however, with a population in 1821 of 52,889, and even against Bath, with 36,811, Exeter (23,479) was small. Coaching services – a dozen or so between Exeter and London, taking nineteen hours for the fastest and twenty-two hours for the mail, and twenty-nine provincial services – seem to have been much on the same scale as at Bath, but there was nothing to generate the volume of traffic, for passengers and goods, which Bristol handled, nor was there any equivalent of the Somerset

Coal wagons made up much of the traffic on the roads of the Bath Trust (p. 121), though this one is probably in the North of England.

Fotomas Index

A mail coach for the West of England in Piccadilly, 1829. William McAdam (p. 136) mentions a service which took him from Wilton, near Salisbury, to Exeter in a day, roughly along the line of the A30.

Fotomas Index

Cheapside, 1823, showing the traffic of late Georgian London.
Museum of London

coal trade on the Bath roads. It is hardly surprising, then, that the Exeter Trust's income from tolls was smaller than Bristol's and smaller per mile of road than the income at Bath.

The Exeter Trust, nevertheless, was much richer than most and its trustees were active and enterprising. In pursuit of ambitious improvements they ran up a large, though soundly managed, debt, and there was plenty of work for surveyors.

The Exeter trustees first became acquainted with the McAdams after they authorized their clerk, on 5th October 1819, 'to apply to Mr McAdam' – presumably J L – 'to find a person properly qualified to inspect the Road from Larkbear to Topsham Bridge to make an Estimate of putting the same into Complete Repair according to Mr McAdam's system, and of the Annual expense of keeping it in repair afterwards.' Mr McAdam sent William, his son, who, on 5th November 1819, was appointed Surveyor of that part of the Topsham road, to put and keep it in repair for 2½ years, being allowed £40 a mile for doing so.

This minor appointment may have been intended as a trial of William's abilities and of McAdam's 'system', for the trustees were dissatisfied with the state of their roads. In 1816 they had let the repairs on contract, some at the unrealistically low rate of £25 a mile, and the upshot, which ought surely to have been foreseen, was that by 1820 they were in 'a most ruinous and alarming condition'.[20] In April 1820 they decided to withhold payment from the contractors for about two-thirds of the trust's mileage and to order a survey of the quantity, quality and value of the stones on the roads. They also resolved that it was 'expedient that some new plan be adopted for the Improvement and maintainance of the Roads of this Trust', and they told the clerk 'to write to Mr McAdam to request him to come to Exeter that he may be consulted on the state of the Trust'. Whilst still in this fierce mood they sacked their surveyor, Hemmingway, for incurring expense without authority 'to a considerable Amount' for repairing and altering Exe Bridge. Hemmingway evidently fell foul of the trustees when they were feeling particularly vindictive – seeking a scapegoat, perhaps, for the results of their own parsimony – and possibly he deserves sympathy.

About a month after this outburst of vigour, a report came in from William McAdam. 'I have inspected the whole of your Roads', he said. 'They are very capable of much Improvement, having at present the appearance of the worn out and neglected Pavement of a decayed Town more than of Roads, and I could not avoid remarking that in three visits to Exeter, and having during the last gone over 150 Miles of Road, I have only seen one Boy on the whole Line, or preparing Materials in Pits, and he was doing no good.'

He went on to say that if the debt of the trust (Table VI) had been 'of a more moderate Amount' he might have advised adding to it 'for the more immediate amendment of the Condition of the Roads', but since it was so great that would be unwise, especially since they could get the same result within their present means, 'only at a more protracted Period'. He then set out the following calculation:

		£
'I take your Tolls at p. Ann [per annum]		10,000
Interest of Debt	£2,710	
Sinking Fund	1,000	
	———	3,710
		£6,290
150 Miles of Road at a Maximum to cost £32 per Mile including the Sub-Survey^s Salaries		4,800
	Surplus £1,490	
Incidentals say	£1,000	
Tresurers Sal^y	80	
	———	1,080
	Farther Surplus	£410 exclusive of Compositions.'

The figure of £32 a mile for repairs, he said, would be found to be about the average of the contracts not suspended – that is, of the more satisfactory ones – and it should be fixed as an 'Average Maximum Cost'. It would then 'during the first Year smooth your Roads so as to allow the Traveller to pass with every facility and if variously applied as the gradual Alteration in the State of the Trust requires, in 3 or 4 Years compleat a great Improvement'.

One contractor had repaired Fire Stone Hill with flint, 'one of the best of Road Materials, but the most improper for making a Hill where Wheels are dragged, the Consequence is that this Hill is a Bed of loose round Flints, very difficult for Horses to drag a Carriage up'. In a typical McAdam passage, William put the blame chiefly on the surveyors:

> 'Here the Contractor has put himself to extra Expence because he knew no better, and your Surveyors being uneducated in the Scientific Principles of Road-making were incapable of directing him – I only mention this as a Case in Point, and with a view to Improvement, not to find fault.'

This led naturally to the observation that the trust's disposable revenue was 'too confined to make experiments', so that 'known and tried Work must be applied under a sufficient Number of Educated Sub-Surveyors who have had the Experience necessary to make the most of Money'. He suggested dividing the roads of the trust into four 'out Sub-Surveyorships' and an 'Exeter or Home Division, which latter should comprize all the Roads of any Consequence for 5 or 6 Miles round the City, and the whole of the London and Mail Lines.' He then explained how he would propose to divide the available money among the sub-surveyors and the measures he would take to make sure it was properly accounted for.

He then asked to be allowed to appoint his own staff, and indicated his intentions towards the 'uneducated' surveyors then in office. 'You have 3 Surveyors,' he said, 'who would be quite Useless to me. If I am to make a Pledge I must be allowed my own Means to redeem it, nor would it be good Policy for the

Commissioners to leave me an Excuse in Case of Non-Improvement, to say that the Issue would have been different had I had my own People. Where Roads have been put under my Charge the old Surveyors have generally been paid six Months Salary and discharged, And I have offered them if they like to go with that Advance and qualify, I would give them Situations, but we never place a Sub-Surveyor in his own Neighbourhood . . . and they always work better and more independently where they are entire Strangers.'

After propounding that principle of personnel management he told the trustees that if it was their desire that he should practically prove his Opinions on the Roads he was ready to do so 'on being allowed a moderately fair Remuneration'. He then explained how he would go about the job:

'I have no Secrets, and it may be asked how will I do so. I should draw from my more finished Roads such Surveyors as had been instructed, and I should send to the Exeter Division (with an Eye over the Rest and an Assistant) one of my steadiest and most respectable Sub-Surveyors. For my own Satisfaction and at my own Expence I would leave my Son constantly at Exeter – my Residence is on this Side of Wilton [near Salisbury], and by the Mail Coach which passes my Door at nine o'Clock in the Morning I am set down at 9 the same Evening in Exeter. I should survey the Work as often as I do other Roads, and at all Times, when my Son reported any difficulty. When I worked the Reading Road I resided at Bristol with my Father which was a greater Distance and no Inconvenience was felt. I conceive that I could have the Surveyors at their Posts in a Fortnight, and the Work begun on all the Roads as soon as Men could be assembled. I am just finishing a Trust between Warminster & Bath and shall have a disposable set of good Workmen to take the head of Gangs elsewhere.'

This report has been discussed in detail because, like the Truro correspondence discussed in Chapter IV, it displays the McAdam system in its fullest vigour. A trust applies to J L

McAdam for advice. He responds, and sends one of his sons who travels the roads and reports on them trenchantly and with supreme self-confidence, offering to put them in order at a price within the means of the trust. The assault on contractors and on 'uneducated surveyors' is fundamental to the whole McAdam approach: so is the request to be allowed to put in surveyors and workmen whom the McAdams have themselves trained, and William's offer to put in his son, at his own expense, is a gesture of respect towards a large, important trust. The dismissed surveyors of the old régime are offered a chance to 'qualify' for inclusion in the mobile work force at the McAdams' disposal, the deployment of which, in this particular case, is explained in detail.

The Macadam Group, as one is tempted, quite unwarrantably, to call it, could by this time offer a complete range of advisory and executive services, from the original consultation with J L McAdam himself through survey, report and recommendations to continuous supervision of the day-to-day tasks of repair and improvement on turnpike roads. Forcefully presented, in a determined thrust for new business, it is no wonder that the package looked attractive to turnpike trustees determined to be up to date and to respond to a demand loudly expressed and influentially supported: better roads for faster travel.

On 2nd May 1820 forty-two trustees met, as the trustees usually did, at the Globe Tavern. They decided to appoint William McAdam General Surveyor and to direct the contractors to give up the roads into his hands. He was to manage them at an expense not greater than the figure he had suggested: £32 per mile/year. His salary, curiously, is not mentioned in the minutes, but it appears from the return of 1825 to have been £420 and was no doubt calculated to include an allowance for William Jr, the son whom William Sr first stationed in Exeter. The sub-surveyors in office were given notice that the trustees intended 'to dispense shortly with their further Services', and about a month later two of them but not, apparently, the third, were paid off as McAdam had recommended with six months' advance salary.

McAdam's appointment marked the beginning of a period of new vigour in the Exeter Trust's affairs. Many more trustees than formerly took to attending meetings and large projects of improvement were put in hand. Sir Thomas Baring Bt, MP, of the banking family, a strong parliamentary supporter of J L McAdam, was one of the trustees, no doubt willing to back strong measures, and one of the most hard-working improvers seems to have been the Treasurer, H M Ellicombe, another convinced macadamite. He was a freeman of Exeter, by profession an attorney, who had been Clerk to the trust as well as Treasurer before the Turnpike Act of 1822 forbade one individual to hold both offices.[21]

During McAdam's early years three improvements needed parliamentary sanction. One covered $4\frac{1}{2}$ miles of new road near Haldon about 10 miles south of Exeter towards Teignmouth. In 1820 a Bill was promoted for a new road towards Plymouth, opened in the autumn of 1822, the old road being 'thrown on the parishes'. In 1821 estimates were sought for 8 or 9 miles of new road towards Okehampton and J L McAdam was asked to survey the line, evidently as a check on the estimates submitted, being thanked for his 'gratuitous services' by twenty-eight trustees, presided over by the Mayor of Exeter, on 6th March 1821. His idea of the cost was £5,479 12s but when William finally announced completion, in July 1824, the cost had reached £6,575 and his report has a defensive tone about it, not only as to the cost, which he attributed partly to a change of line, but also as to the time taken. He begged the trustees 'to have the goodness to remember that the work has from first to last been carried on in this valley of deep running clay during two of the wettest years ever remembered'. It had been his interest as well as theirs, he pointed out, to finish the work, 'but I have not thought of my own Interest, nor of forcing their work when I saw that it would only be burying money in wet soil'.

Early in 1825 the trustees decided to apply for a new Act, contemplating an outlay of £10,000 on improvements, and it was passed in March 1826. They spent in fact far more than £10,000 during the years that followed and the balance of the bonded debt mounted from £54,650 in 1823 to £92,300 in 1829 after £16,900 had been borrowed as an 'additional sum . . . appropriated to the New Road Formations'. One of these, probably, was a stretch of the Barnstaple road, about 8 miles long, between Copplestone and Eggesford Bridge on the modern A377, which was built or rebuilt as part of an ambitious scheme undertaken in co-operation with the Barnstaple Trust, another large trust with 68 miles of road, to which William McAdam had been appointed surveyor in 1821. He became surveyor at Totnes in 1822, at Plymouth in 1823, and at Kingsbridge for three years in 1824 at the unusually high salary of £397 10s, so that in the mid-twenties McAdam influence was as widespread round Exeter, where Christopher also may by this time have been starting to collect surveyorships, as round Bath and Bristol.[22] It was not, however, destined to last so long.

By this time William's repairs were becoming more expensive. In May 1825 he asked the Exeter trustees for £37 a mile/year instead of £32, justifying his application 'solely from the present improved State of the Trust Funds, and the altered State of the Country, as to Expenditure in travelling, encreased Trafic, and the comparatively better condition expected in all Roads, and particularly looked for in an Important Trust like yours where Means exists'. Towards the end of the same year he was allowed £40 a month for six months on the new road towards Okehampton, and in May 1827 £500 for one year.

In April 1826, under the terms of the new Act, William McAdam was again elected Surveyor 'at the usual Salary' and the Thanks of the Meeting were given him 'for his judicious Attention to the Duties of his Office'. His deputy by now was his son Christopher, not William Jr, who joined his grandfather at Bath in 1826, and in October 1827 an agreement was negotiated by which father and son were appointed joint and separate surveyors 'for the term of five years if Mr William McAdam shall so long live . . . at the former salary of £420'.

In 1825 or thereabouts the McAdams' standing in Exeter

probably reached its peak. It may not even then have been so securely based as it appeared to be, and in May 1827 William McAdam made a pre-emptive strike against an attack which he must have known for some time had been building up against him, gathering force, no doubt, from the rising costs of repairs. On 1st May he 'laid before this Meeting a report upon the proceedings of this trust as effecting [*sic*] the Expenditure thereupon so far as it is applied to the maintenance of roads' and asked for 'an Investigation by the trustees of the roads generally . . . and his Management thereof and outlay thereon'. A committee to make the investigation was at once appointed, open to all trustees.

The Committee spent five months investigating and deliberating, and a very thorough job they did. Their report, submitted on 30th October, ran to seven sheets of foolscap, supported by a five-page report from Ellicombe as Treasurer and various other documents, including a copy of a five-and-a-half-page letter from J L McAdam to the Bristol trustees on the virtues of his system. The committee went at length into McAdam's system of management and financial control and were entirely satisfied. They found that Christopher McAdam lived in Exeter and was employed about nine months of the year 'in superintending the Work done upon the Roads . . . and in examining the Accounts'. The General Surveyor, they said, 'comes here occasionally when he remains several Days and goes over the greater part of the Roads of the Trust in the course of the Year.' Then they turned to 'objections . . . made by one of the Committee to an adherence to the present system':

'Firstly, That under no System adopted by this Trust for the last ten years had the Road in his particular District been in so bad a state of repair.
'Secondly, That Mr McAdam is the Examiner and Surveyor of his own Work which is incompatible with General Economy and is contrary to the System observed in every other public Expenditure in this County.'

They dealt with the first objection by observing 'that the General state of the Repair of the roads belonging to this great and extensive Trust was the object to which their attention was principally directed', but that if there had been any complaint about a particular district 'to which the General Surveyor had not been able to give a clear and satisfactory answer it would have been incumbent on the Trustees to take care that such a complaint was redressed and in that event to record their sense of the neglect in question'. They recorded no sense of any neglect.

The second objection, carrying an innuendo against McAdam's integrity, they rejected utterly. 'He has no other Interest to promote,' they said, 'respecting the work performed upon the Roads than the welfare of the Trust, . . . he does not participate in any manner whatever in the Money expended (which in fact does not even pass through his hands) . . . [and] the regular discharge of his Duty unbiassed by any sinister motive calls upon him to see that the Trust Money is not paid to the parties claiming it unless a just and fair supply of Materials and a skilful application of adequate and satisfactory labour show that they are entitled to receive it.' They also pointed out that every trustee could judge for himself 'whether all this be performed or not' by looking at the accounts made up every fortnight at the Road Office and laid before the trustees at the end of each year. Even then, the committee had not finished with McAdam's detractors. They looked at figures of turnpike expenditure laid before Parliament, varying from £500 a mile in Middlesex to £21 in Westmorland, and discovered 'that in the Number of 40 Counties mentioned therein this great County stands so low as No 28 and incurs an Average Expenditure of only £37 per mile which appears to them to afford Ground for concluding that the Mileage Expenditure upon this Trust is as low as can be reasonably expected'. They reported also that several trusts which had abandoned McAdam's system had regretted doing so, 'and that, after having experienced the effects of a departure from that System the Bristol Trust had again adopted it'. That must have been highly gratifying to J L McAdam, and amongst the papers supporting their report the Exeter committee included the copy already mentioned (see p. 140) of a letter written by him in June 1827 'to the Bristol Trust on calling for his assistance after having abandoned his System'.

To a report of this kind there could be only one conclusion :

'The Committee conclude for the reasons abovementioned that the operations of this Trust have been conducted more beneficially to the Public, and with greater Security to its Creditors under the System of the present Surveyor General than at any former period; and are of opinion that in the absence of all Evidence to shew any Misapplication of its Funds or Dereliction of Duty on the part of that Officer, it would be injurious to the Public, and unjust to the Creditors of this Trust, to dissolve a System which is so advantageous in its results to both those parties.'

For William and Christopher McAdam the Committee's findings must have seemed almost too good to be true, allowing no word of criticism either of them or their 'system' to stand. In the wake of the report, at the meeting at which it was presented, came their election for five years as joint General Surveyors (see p. 139). The same meeting ordered the raising of £6,000 for the new road (see p. 139) between Crediton and Eggesford Bridge and asked McAdam for an estimate for widening the road from Exeter to Exmouth.

After another five years of very active – and expensive – road building and improvement the trustees came again to the question of continuing their surveyors in office. Seventy-one trustees met at the Globe Tavern on 28th March 1832 to vote on a motion to reappoint them jointly and severally for five years, and thereafter at six months' notice. The terms of the motion – 'it appears to this Meeting that the System upon which the Roads of the Exeter Trust have been repaired for the last ten years under the Superintendence of Mr William and Mr Christopher McAdam has tended greatly to improve the Roads, and benefit every description of travelling . . .' – were complimentary, and the motion was passed. It was not passed, however, without a considerable body of dissent. Fifty-three trustees voted for it: eighteen against.

In 1832, after the borrowing of £6,000 for improvements at the North Entrance to the town, the Trust's bonded debt reached £101,800, a very high figure indeed, and some trustees became alarmed. Mr Sillifant, who had proposed the re-election of the McAdams two years earlier, gave notice in March 1834, at a meeting attended by fifty-three trustees, that he would move to cease borrowing until the debt had been reduced to £100,000. When the motion came on, in June, it had the support of the Treasurer, who said that the debt, then standing at nearly £110,000, had become 'largely disproportionate to the Annual Income'. The trustees agreed, with two dissentients, that no more should be borrowed until the total had been brought down to Sillifant's figure of £100,000.

This move in itself need have presented no threat to the Surveyors General, whose concern was not with debt, but expenditure. The mood of the times, however, was scarcely propitious towards that, either, and it is probably not accidental that in August 1835 Captain Hamlyn gave notice of motion to consider replacing the existing system of repairing the roads with 'a different Contract System': that is, of returning to something like the system in force before William McAdam was appointed in 1820. Hamlyn never seems to have put his motion, as he intended, to the Annual General Meeting, but it was ominous that he should have gone so far as giving notice to do so.

In February 1836 William McAdam died. He was in Scotland at the time, in his house at Ballochmorrie, where for some years he had been building, suggesting that he was comfortably off and also, perhaps, that he was losing interest in his work in the South West of England. In any case, his death can hardly have strengthened the position of Christopher, left as sole General Surveyor and due for re-election – or dismissal – in 1837.

In March of that year Mr Buller, of the ancient county family to which Redvers Henry Buller (1839–1908), Britain's remarkably unsuccessful commander in the early part of the Boer War, also belonged, gave notice of motion to appoint a Committee to consider what steps should be taken in consequence of Mr McAdam's contract terminating in March 1838.* This time the intention was carried out. In June a meeting at which twenty-six trustees were present appointed a committee of twelve, with Buller in the Chair and Hamlyn a member, to do exactly what Buller proposed. Buller, as if his meaning were not already clear, gave notice that at the next meeting he should move 'that Notice be given to Mr McAdam that his contract terminates at Michaelmas 1837' and at a meeting of thirty-four trustees in August 1837, the Clerk was duly instructed to give Christopher McAdam notice that his contract would be ended at Lady Day.

This time Christopher put up no fight. The trustees, having decided on circular motion, voted to go back to the system of putting repairs out to contract which had served them so ill between 1816 and 1820 and then bade Christopher a ceremonious farewell. They recorded 'their sense of the great Improvement effected in the roads of this extensive Trust since the Year 1818 when Mr McAdam's Father first undertook to superintend their repair and management'. They felt assured 'that the Agricultural Commercial and Manufacturing Interests of this Neighbourhood, highly appreciate the increased comfort and safety, as well as the profit, resulting from the Ability and attention with which Mr McAdam here introduced and carried out, a system for which the whole Country has acknowledged its obligation to his family'. It was gratifying for them, they said, to add 'their high sense of the Strict Integrity with which Mr Christopher McAdam has for several Years discharged the duties of his Office, and the unvarying Courtesy which has on all occasions accompanied his intercourse with the Members of this Trust'. Christopher, whatever his private thoughts may have been, maintained his Courtesy unvaryingly to the end and

* This date presumably allows for six months' notice after the ending of the five-year agreement of 1827.

replied in suitably emollient terms on 13th October 1837. Thus placidly, and with all due decorum, the McAdam family ended nearly twenty years of service to the Exeter Turnpike Trust.

Christopher himself withdrew steadily from surveyorships over the next three or four years. His appointment at Truro, which he had shared with William Sr until 1833 and then undertaken alone for three years, he gave up in 1837, and he gave up the Teignmouth–Dawlish surveyorship, to the new Exeter surveyor, in 1838. He gave up two surveyorships in Cornwall, at Halesworthy and Launceston, in 1840 and 1841, but the surveyorship of the Countess Wear Bridge in Exeter he kept, perhaps for sentimental reasons, until 1846.

J L McAdam & Grandson at Bath, 1826–1861:

When Christopher was removed from his post at Exeter, William Jr had been at Bath for eleven years, and he was to remain for almost a quarter of a century. He held also, in succession to his father, two surveyorships in Somerset – at Black Dog and Frome – and one – Southampton South District – in Hampshire. William did not acquire many appointments, but those he had, he held, conveying an impression of quiet competence, never over-extended.

When William and his grandfather took up their duties at Bath, in the Spring of 1826, the trustees, like those at Exeter, were full of enthusiasm for improving their roads to suit the bustling traffic of the day, and they were better placed to do so, for they had both a larger income per mile of road and a smaller bonded debt. Before Christmas 1826 they asked their new surveyors to inspect the roads with a view to avoiding or lowering hills with a gradient steeper than one in twelve and to improving the roads generally by widening or shortening them. If they agreed with the McAdams' recommendations they would seek a renewal of their local Act to give them the powers they would need.

BATH TURN

INCOME.	First District.			Second District.			Third District.		
	£	s.	d.	£	s.	d.	£	s.	d.
Balance in the Treasurer's hands...	...	...	...	3,432	10	9¾	...	...	...
Revenue received from Tolls ...	3,167	8	6	3,701	17	0	3,893	1	2
Parish Composition in lieu of Statute Duty	...	...	...	...	...	...	...	...	...
Estimated value of Statute Duty performed	...	...	...	...	...	...	...	...	...
Revenue from Fines	...	...	...	...	...	...	...	...	
Revenue from Incidental Receipts...	132	0	0	70	0	0	97	10	0
Balance due to the Treasurers ..	955	5	1½	...	...	...	3,535	15	0
	4,254	13	7½	7,204	7	9¾	7,526	6	2

DEBTS.	First District.			Second District.			Third District.		
	£	s.	d.	£	s.	d.	£	s.	d.
Bonded or Mortgage Debt	15,350	0	0	12,600	0	0	15,879	0	0
Floating Debt	48	9	3	67	9	1	93	1	1
Unpaid Interest	...	...	...	...	...	...	...	...	...
Balance due to the Treasurers ..	955	5	1½	...	...	...	3,535	15	0
	16,353	14	4½	12,667	9	1	19,507	16	1

At the General Annual Meeting of the Trustees of the Bath Turnpike Roads, held at the Guildhall, in Bath, on Saturday, the 20th March, 1841, in pursuance of the Act of the 3rd and 4th William IV., c. 80, the above General Statement was submitted to, and approved by, the Trustees assembled at such Meeting.

JOHN STONE, *Chairman.*

PIKE TRUST

Turnpike Trust, in the County of Somerset, between the 31st day of December, 1840.

EXPENDITURE.	First District.			Second District.			Third District.		
	£	s.	d.	£	s.	d.	£	s.	d.
Balance due to the Treasurers ..	421	5	5½	...	...	...	2,562	6	1
Manual Labour	348	16	10	455	7	11	645	15	2
Team Labour	794	15	11	818	5	8	735	18	8
Materials for Surface Repairs ...	1,003	0	2	1,311	11	10	1,572	8	10
Land purchased	...	...	...	...	...	...	...	...	...
Damage done in obtaining Materials	38	2	4	49	3	8	82	12	9
Tradesmen's Bills	84	5	9	62	1	5	81	13	10
Salaries	327	1	4	327	1	4	327	1	4
Law Charges	12	1	3	37	3	4	12	1	3
Interest of Debt...	695	5	0	571	10	0	780	19	4
Improvements	105	14	3	271	2	10	293	6	7
Debts paid off	400	0	0	400	0	0	400	0	0
Incidental Expenses	24	5	4	24	2	4	32	2	4
Statute Duty performed, estimated value	...	...	...	...	...	...	...	...	...
Balance due to the Trust... ...	...	...	...	2,876	17	5¼	...	...	...
	4,254	13	7½	7,204	7	9¾	7,526	6	2

Rate of Interest	ARREARS OF INCOME.	First District.			Second District.			Third District.		
		£	s.	d.	£	s.	d.	£	s.	d.
4½ per cent.	Arrears of Tolls for the current year, and Compositions	...	...	...	...	...	...	131	5	0
	Arrears of Parish Composition ditto	...	...	...	...	...	...	...	...	...
	Arrears of any other Receipts, viz.—									
	County Bridges ...	44	0	0	15	0	0	32	0	0
	Balance due to the Trust	...	...	...	2,876	17	5¼	...	...	...
		44	0	0	2,891	17	5¾	163	5	0

PHILIP GEORGE, Clerk, Guildhall, Bath.
Messrs. TUGWELL, MACKENZIE, and CLUTTERBUCK, Treasurers, High Street, Bath.
THOMAS HARVEY, Sub-Treasurer, Northumberland Buildings, Bath.
WILLIAM McADAM, General Surveyor, 18, Westgate Buildings, Bath.
RICHARD DOSWELL, Holloway, Bath, ⎫
GEORGE DOSWELL, Lambridge, Bath, ⎬ Sub-Surveyors.
THOMAS PERRY, 18, Westgate Buildings, Bath, ⎭

DISTRICT AND SITUATION OF THE IMPROVEMENT.   Sum paid.

	£	s.	d.

FIRST DISTRICT.

	£	s.	d.
Bannerdown Hill	47	14	0
Toll Houses	6	17	8
Repairing Walls	1	13	3
Fences	32	4	8
Drains	9	10	1
Incidents	4	8	8
Watering Account	3	5	11

SECOND DISTRICT.

	£	s.	d.
Repairing Walls	16	4	0
Weighing Engines	104	19	8
Holloway Arch Improvement	143	9	2
Wall, Macaulay Buildings	6	10	0

THIRD DISTRICT.

	£	s.	d.
Repairing Walls	20	8	6
Depôts for Stone	6	19	1
Widening at Twerton	30	0	0
Repairing Fences	22	1	3
Toll Houses	11	0	5
Cross Post and Mead Lane	128	16	5
Pumps	7	14	10
Land Purchased	3	0	0
Railway Arch, Lower Bristol Road	55	2	7
Improving Road opposite Newton Lodge	8	3	6
	£670	3	8

OBSERVATIONS.

The Road and Land having slipped away, it was found necessary to divert the road on
  to the solid ground for an extent of about 300 yards.
General Repairs.
Resetting Coping mischievously destroyed by passengers.
Repairs to Thorn Hedges now given up.
Drain to the back of Foot-path in Shockerwick new Road.

Extra posts for Pumps, and altering a wall to get at the river.

Re-setting Coping and Edging to a Foot-path.
A new Engine, Combe Down Gate, and repairs to that at Brass Knocker.
Lowering the road and Foot-path, and erecting a Wall against the railway premises on
  the south side of Holloway Arch, £25 was paid towards this outlay by the Company.
Towards erecting a high Wall which supports the Road, and which had given way in
  consequence of age, and alterations made by the tenant.

Re-setting Coping and repairing Locksbrook Bridge Walls.
Repairing Walls of Depôts.
Widening the Road north side under Mr. Wilkins' Wood, Twerton flat.
Eight Mile Hill, and High Littleton.
General Repairs.
Widening the Road on both sides at the Weston end of Cross Post new Road.
Turner, for a Jig Pump and repairs.
Land given up by setting back Iron Railings in front of Whiting's house, Twerton.
Raising the Lower Bristol Road, in consequence of the erection of Holloway Arch.
Taking off a hump in the Road, widening it, and lowering the Footpath.

The trustees were already aware of an ambitious proposal, supported by trusts at Bradford-on-Avon, Melksham and Devizes, for a completely new road about 7¾ miles long. It would link Melksham to Bath, and therefore London to Bath, by a route to the south of the existing one and much straighter. Telford had been called in, which cannot have been welcome news to the McAdams and may betray the influence of Benjamin Wingrove, who was surveyor at Bradford in 1834 and may have gone there direct from Bath in 1826.

There is no evidence to show how much time, if any, Telford spent on the project, but the plan of his intended route, which survives, looks very much as if the great man had done little more than rule a couple of straight lines on a map and say 'So be it'.[23] The Bath trustees nevertheless were alarmed, presumably by the competition the new road would offer to their own routes from London by Chippenham (A4) and by Devizes. Scobell and other trustees met the trustees from Wiltshire to discuss the proposal and, having discussed it, decided they didn't like it. In February 1827 Scobell and others were instructed 'to watch and, if necessary, to oppose an application to Parliament by the Trustees of the Melksham and Bradford Roads for a new Line of Road from Melksham to Bath'.

They were not purely destructive in their opposition. They asked J L McAdam to meet representatives of the other side. Having done so he prepared a plan of his own, less ambitious than Telford's but more carefully worked out. He proposed to provide a connection between the Chippenham Road and the Devizes road, about three miles long, which would cut out or ease two serious hills: Shayler's Hill on the road to Chippenham and, far more severe, Kingsdown Hill on the Devizes road. On that road, said the trustees when they published McAdam's plan, it would effect 'the enormous reduction of 194 feet in the present Hill' and the gradient, which in the steepest part was one in nine, would nowhere be greater than one in thirty-six on the new road. The saving in time, they suggested, might be about five minutes towards London and ten or twelve minutes away from it: an interesting illustration of the importance people were beginning to attach to exact timing and to savings of a few minutes. 'It is . . . highly probable,' the trustees concluded, 'that his improvement would induce others on an extensive scale on both lines of Road, and thereby greatly facilitate the intercourse between large and populous portions of the Empire, and particularly between the Metropolis and the south of Ireland.'[24]

McAdam won. Telford's plan was abandoned and McAdam's became the principal feature of the recommendations he put to the trustees in June 1827. He was most anxious to see it carried out, describing the road from Melksham through Bath and then towards Wells (A39) as the most important of any in the kingdom 'distant one hundred miles from London' and adding 'And if I were called upon to name the most inconvenient, steep, and dangerous great road in England, I should say it was *that* line.' He warned the trustees that if they did not make the improvement, another trust might be set up to do it, which would be a great reflection on them. He recommended also various other improvements on the two roads from Bath towards Wells (A39 and A367, through Radstock), and suggested borrowing £30,000 for all these purposes which required altogether some 20 miles of new road. The trust was rich enough to do it, he suggested, and he warned the trustees that there was otherwise a danger of serious competition:

'The Revenue of the Bath Trust is so ample as to justify the public expectation, that every measure will be adopted on these roads, that can benefit the city and county adjacent, and that all the facilities which the country will admit will be given, in order to continue the communication between London and the West of England, in its present course through the city of Bath, so as in a great measure to preclude the necessity and diminish the chance of that communication being opened by way of the Vale of the White Horse to Bristol, and from thence by the level line to Bridgwater.'[25]

The line he describes through the Vale of the White Horse is the line he had proposed for a road to complement the railway projected, but never built, in 1824 (see p. 62), and it is very much the line chosen by I K Brunel for the Great Western Railway, though he varied it to bring the railway through Bath instead of going direct to Bristol.

The proposal for a new Act stirred some of the trustees into energetic manoeuvring to get it dropped, led to lengthy negotiations with property owners, and displeased at least one neighbouring trust. The terms were not finally settled until February 1829 and by then the expenditure contemplated had risen to more than £38,000. Captain Scobell and three others were authorized 'to attend Mr George [the clerk] in the prosecution of the Bill through . . . Parliament' and to require McAdam's assistance when it was needed. George was granted £500 for the purposes of the Bill and in May £500 more. Even that was probably not enough, for in June the minutes complain of 'vexatious opposition' to the proposed table of tolls and to certain projected improvements, conducted on behalf of the Coach Masters by Mr Boord and by Mr Allen for the parish of Walcot, which would add still more to the cost of the Act.

At last, on 1st October 1829, the trustees held their first meeting under the new Act, and the improvements could go ahead. McAdam's plan for the Devizes road was put into effect and today its results can be seen in a curve of A4 just on the Bath side of the village of Box and in about a mile and a half of A365 as it leaves A4 in Box and passes under the slope of Kingsdown Hill to rejoin the old line, over the top, making for Melksham.

The McAdam partnership was in high favour. In June 1830 William was made joint Surveyor with his grandfather, rather than merely deputy. In June 1831, after a committee had reported on the state of the roads and the General Surveyor's expenditure over the previous five years, the trustees presented to them 'their strongest (and unanimous) thanks for the zeal, efficiency and integrity with which they have executed the office of General Surveyor and so amply redeemed every expectation'. They were reappointed for five years and then 'until superseded' at £600 a year 'for the joint and separate duties of the office as the case may be', one of them to be 'usually superintending the Roads and Accounts' and the appointment was to continue after the death of either of them.

They found themselves, soon afterwards, caught up in controversy between the Bath trustees and the trustees of Black Dog, who in July 1832 announced their intention of going to Parliament next session for an Act to build about 12½ miles of new and adapted road to connect Warminster with Bath along the valley of the Avon instead of going over the very steep road, the only one then existing, across Claverton Down. When the Bath trustees heard of the proposal they did not like it, for it would divert traffic from their own gate at the foot of the Down, known as Brass Knocker. In October 1832 a committee produced a rival scheme. It only required six to seven miles of new road, but it still went over the Down, and the only point in its favour – admittedly a strong one – was that Black Dog's scheme would cost 75 per cent more. On the Bath trustees' own computation, however, taking account (though they do not say how) of 'the comparative joint effect of length, elevation and planes', the more expensive line would be 8½ per cent more efficient. It seems a very sophisticated calculation for its day, and it is a pity the details do not survive. The Black Dog proposal went through, and to this day is the basis of some twelve miles of A36(T).

The surveyor to the Black Dog trust was William McAdam Sr, so that he found himself at professional loggerheads with his father and his son. The son, William Jr, was probably on both sides at once, for he was certainly associated with his father at Black Dog in 1834 [26] and may well have been there two years earlier. That he should have contrived to remain on good terms with both trusts reinforces the impression, gained from other aspects of his career, of combined tact and ability.

The Black Dog project must have been among the last considerable building schemes put forward before the shadow of the railways began to fall across the turnpike roads of Wiltshire and Somerset. The first the Bath trustees heard of any

railway, officially, seems to have been on 1st February 1834
when Mr Hooper, a Bristol surveyor, waited on them with
'plan and section of the intended Railway from Bristol to Bath'.
Hooper they sent away, demanding instead 'Mr Brunell or
some person fully authorized to act on behalf of the Promoters',
and a week later Brunel and others duly attended.

The Great Western Railway's Act was not passed until 31st
August 1835 and construction works made little impact on the
territory of the Bath Trust until 1837. In the interval, the
improvements authorized by the trust's Act of 1829 went
forward and the trustees, in 1835, once again recorded their
satisfaction with their two surveyors.

J L McAdam, in his late seventies and living at Hoddesdon,
was in the habit of migrating to Scotland in the summer of
every year, travelling in a two-horse carriage followed by a
pony and a Newfoundland dog. The purpose of the dog was to
see that the pony did not dawdle or stray and the purpose of the
pony was to provide McAdam with a mount if he wished to go
sightseeing while on the road, or perhaps use his fowling
piece.[27] He was accustomed to start back from these visits in
the autumn and in 1836 was preparing to do so when at 3 am
on 26th November 1836, at Moffat in Ayrshire, he died, as
his son James recorded, 'without pain or even complaint for he
died of old age after a useful, long and honourable life'.[28]

'Since the operations on the railroad commenced,' reported
William McAdam in 1839, 'the increase and destructive nature
of the travelling has been such as to baffle anticipation or
estimate.' Moreover the winters of 1838–1839 and 1839–1840
were severe, and the combined result of bad weather and railway building, between 1837 and 1841, was to carry the
surveyor's expenditure on repairs far beyond his estimate:
something which had never happened between 1826 and 1836
'notwithstanding the gradual increase in travelling'. There had
if anything been a drop in expenditure, for William had
dropped his estimate in 1832–1833 and had relinquished £200
of salary in 1837, after his father died. The normal level of
expenditure was below £7,500, but while the railway work

was at its height, in 1839–1840, it was running at nearly £9,000 a year (see Table V).

After breaking up the turnpike roads with their construction traffic, the railways would rapidly ruin the trusts with their competition, as William McAdam was well aware. 'In about two years,' he said in his report for 1837, 'it may be expected the line of railway will be completed between Bristol and Bath, by which, at a rough calculation, this Trust will lose in coaches alone £889 per annum. In three years from this time, we may presume the whole line will be opened, by which the Trust will sustain a further loss by coaches of £882, together £1,771.' In fact, as the table shows, the figures tended inexorably downward far below the level forecast by McAdam, and in 1851 the passing of the Bath Improvement Act made matters worse by extinguishing all tolls within the city. With the declining income, caused by what the trustees in 1845 called 'the present diminished State of Conveyance or Traffic on the public roads', the sums spent on repairs likewise declined and the enthusiasm for improvement of the twenties evaporated, giving place to anxious consideration of 'the present Manner and Costs by which the repairs of the Roads are effected'.

This gathering gloom need not be explored in detail. William McAdam remained Surveyor to the Bath Trust until he died, on 29th July 1861, at The Park, Bath. The glory of his grandfather's name was still sufficient to earn him a brief obituary in *The Gentleman's Magazine*, which called him 'a man of great talents as an engineer'. He was fifty-eight: a man who had outlived his era.

*   *   *

During the late 1830s, after the death in 1836 both of J L McAdam and of William Sr, the members of the McAdam family, with one exception, dwindled rapidly in influence, for by 1841 only six surveyorships remained in the hands of Loudon (Bristol), Christopher (Countess Wear Bridge), and the younger William (Bath, Black Dog, Frome, and Southampton – South District). James McAdam, on the other hand, succeeded to his father's position as the nation's leading authority on turnpike roads. He was surveyor to thirty-seven turnpike trusts, including the Metropolis Trust, in 1841, and during the forties he added two more, so that by the time he died, in 1852, he was still much the greatest pluralist that the turnpike world had ever seen. It was his misfortune, not his fault, that turnpike roads by the mid-century were far gone in decline.[29]

REFERENCES

1. The trustees' minute books of the Bath Turnpike Trust are in the Somerset Record Office at Taunton, H/10 D/T/ba 10–12. The Exeter trustees' minute books are in the Devon Record Office at Exeter, ETT 2 4–5. I indicate in the text where I have drawn on these sources and have not usually cited them in separate footnotes.

2. Petition to Parliament, D/T/ba 11.

3. Cary's *New Itinerary* 1828, 'A List of Provincial Stage Coaches', 'London Stage Coach Directory', pages not numbered.

4. B J Buchanan, 'Turnpike Roads in a Regional Economy'; Bath Minutes generally and 10xi1810, 4iv and 19xii12; 6iii and 20iii13; 7i15 and printed address 'To the Trustees of the Bath Turnpike Roads'; 7x15; 7xii16, all in D/T/ba 10. Report of the SC on McAdam's Petition 1823, 67.

5. SC on McAdam's Petition, 1823, 53–4.

6. 'Report of John Loudon McAdam', D/T/ba 10.

7. J L McAdam to Bath trustees 15iii17, as (5) 97.

8. As (5) 51. See also Bath minutes which show 109 trustees present, not 107 as in the Report of the SC on McAdam's petition.

9. As (5), Wingrove's evidence, 70.

10. As (5) 62 63.

11. As (5) 63 64.

12. As (5) 63–4.

13. As (5) 69; Wells Trust papers in Somerset Record Office, D/T/wel; papers of various trusts in Wiltshire Record Office, Trowbridge, A/3/7/2/4 7 29.

14. As (5) 96; see also Bath minutes 29x 5xi25, D/T/ba 11.

15. Scobell's appointment, *Navy List*, John Murray, relevant years. Employment of Commanders, Michael Lewis, *The Navy in Transition* 84.

16. Devereux, *John Loudon McAdam* 128.

17. Wilts RO A/3/7/2/16.
18. Shepton Mallet trustees' minutes 3 11viii, 6x23; 1xi24; 4iv 1v 12vi25; 4viii28. Somerset RO D/T/sm 3 4.
19. J L McAdam, *Management of Trusts* 4.
20. As (5) 28.
21. Rowe & Jackson, *Exeter Freemen*; see also Treasurer's Report in ETT2/5 104.
22. Kingsbridge salary – Return of Salaries and Gratuities 1818–25, 6.
23. Telford's plan is in the Wiltshire RO, A/3/16/1/19.
24. Plan and Description of McAdam's proposal, Somerset RO D/T/ba 25.
25. Printed Report 'To the Trustees for the Care of the Bath District of Turnpike Roads' 9iv27, Somerset RO D/T/ba 11.
26. Abstract of Statements of Income and Expenditure of Turnpike Trusts 1836, BPP 1836 (2) XLVII 297.
27. Devereux 130.
28. Sir Jas McA to the Duke of Richmond 26xi36, Goodwood Mss 1874/861; Sir Jas to Mrs Saunders (his sister) 26xi36, quoted by Devereux 148.
29. Based on Turnpike Trust Abstracts as (26).

VII

Zenith and Decline:
James McAdam 1827–1852

Zenith: the Metropolis Roads, 1827–1833:

JAMES MCADAM, in his time and in his temperament, belonged to the generation of reformers who took the administration of 18th-century England by the scruff of its neck and shook it into the beginnings of its Victorian shape. Not for him the part-time amateurism, combined with patronage, of the old order. He believed in full-time professional competence and financial control founded on accurate facts tirelessly collected, tabulated and assimilated, as for many years he personally assembled and set forth the facts, annually presented to Parliament from 1836 onward, about the finances of turnpike trusts in England and Wales.[1]

It is easy to conceive of James, in other circumstances, joining that great school of Victorian administrators, the

Indian Civil Service, as many Scots did, and he had much in common with younger contemporaries, some with Indian experience, who worked for administrative reform and the abolition of patronage. Charles Shaw-Lefèvre, later Lord Eversley (1794–1888), whom he certainly knew and worked with, was one. Others whom he may or may not have met were Lord Macaulay (1800–1859), Sir Edwin Chadwick (1800–1890), Sir Charles Trevelyan (1807–1886) – like Macaulay, at one time an Indian official – and Stafford Northcote, Lord Iddesleigh (1818–1887).

In 1839, James pointed to the parish 'unions', created under the Poor Law Act of 1834, of which Chadwick was a principal architect, as an example of the kind of organization desirable for turnpike trusts. In fact Chadwick may have been influenced by the example of the trusts, for 'consolidation' was a principle long advocated by James, his father and other authorities, and as far back as 1826 it had been applied to trusts in London. In that year, under the terms of the Metropolis Roads Act (7 Geo IV c 142), fourteen trusts, managing between them about 131 miles of the busiest thoroughfares in the world, were consolidated under a newly created body, the Commissioners of the Metropolis Turnpike Roads North of the Thames. James became their Surveyor-General. No appointment, probably, could have been more to his taste, and the pathway to distinction opened before him.

The project for consolidating London trusts had by 1826 long been in agitation, for turnpike gates, irritating everywhere, were infuriating in the largest city in the world. The grievance there was more than a matter of farmers' carts and animals obstructed on the outskirts of a country town, or of coaches and waggons hindered on the 'great roads'. The area covered by the trusts – the 'Metropolis north of the Thames' – stretched a dozen miles or so from North to South and about the same from East to West. Its population by 1826 was over a million and it was growing rapidly. It generated very heavy internal traffic, especially omnibuses; it drew in vast quantities of supplies including animals on the hoof; it was the focal point of

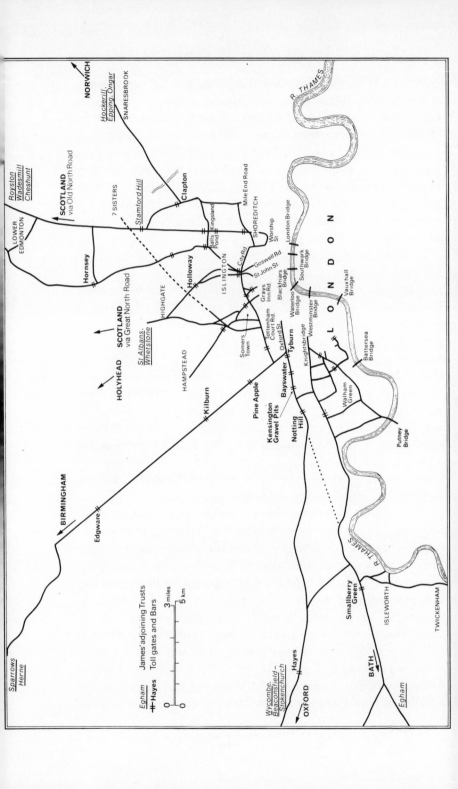

R. THAMES

NORWICH

Hockerill,
Epping-Ongar   SNARESBROOK

Royston
Wadesmill
Cheshunt

SCOTLAND
via Old North Road   7 SISTERS   Stamford Hill   Clapton

LOWER
EDMONTON   Mile End Road

Hornsey   Kingsland   SHOREDITCH   Worship
                                                      St.   London Bridge

Balls
Pond

HIGHGATE   Holloway   ISLINGTON   City Rd.   Southwark
                                     St.John St.   Bridge
SCOTLAND   Goswell Rd.
via Great North Road   Blackfriars
                                     Grays   Bridge   L O N D O N
St Albans: HAMPSTEAD   Inn Rd.   Waterloo   Vauxhall
Whetstone   Somers   Bridge   Bridge
                    Tottenham   Town   Westminster
HOLYHEAD   Court Rd.   Bridge
                                     Oxford St.
                    Tyburn   Battersea
Kilburn   Bridge
                    Bayswater   Knightsbridge

Pine Apple
                    Kensington   Walham   Putney
BIRMINGHAM   Gravel Pits   Notting   Green   Bridge
                                        Hill

Edgware   James' adjoining Trusts
                 Toll gates and Bars
Egham
    ━━┼━━ Hayes   Smallberry   R. THAMES
                                              Green   ISLEWORTH

         0   3 miles   TWICKENHAM
         ┌──┬──┬──┬──┐
         0   5 km   BATH

Wycombe:
Beaconsfield-
Stokenchurch   Hayes   Egham
                 OXFORD

Sparrows
Herne

the Royal Mail and of all kinds of long-distance services for passengers and goods; and it was heavily obstructed throughout, not only on the outskirts but on busy streets (fairly recently country roads) such as Old Street near the City and Marylebone Road in the West End. As late as the mid-fifties, when the Metropolis Roads commissioners and other authorities had done away with many gates, there were still reported to be 87 within four miles of Charing Cross, counting only those North of the Thames.[2]

'It is not the sum of money paid at them that is the grievance,' said J L McAdam in 1825, and his words carry conviction, 'but the continual stoppage. In the throng and middle part of the day, you will see five or six carriages stopt at a time at a gate, waiting till each has settled their account. I reckon the inconvenience to be much greater than the expense; it is such as never would have been submitted to by the people of this country, if they had not got used to it.'[3]

Inconvenience and expense were not all. There was the even more important matter of the state and improvement of the roads themselves. Roads in and about London were notoriously bad, killing coach-horses (see p. 17) at a scandalous rate. 'The gravel of which the roads round London are formed,' wrote J L McAdam, discussing road-making materials, 'is the worst; because it is mixed with a large portion of clay, and because the component parts of the gravel are round, and want the angular points of contact, by which broken stone unites, and forms a solid body; the loose state of the roads near London, is a consequence of this quality in the material, and of the entire neglect, or ignorance of the method of amending it.'[4]

There was no reason why the London roads should not be as good as any others if the management was improved. McAdam, along with others, complained of 'the division of the roads into so many small Trusts, which precludes the possibility of any extended plan of operations, for the benefit of the whole'. If the London trusts were consolidated, management could be brought on to a uniform plan – McAdam's, naturally – so that the best possible use could be made of local gravel. Economies of scale would make possible the importation of better materials, such as granite, from distant parts of the country. Altogether, the roads round the Metropolis might be rendered 'a pattern for the kingdom'.

These are the words of the Select Committee of 1819, which heartily approved the idea of consolidation. So did the Select Committees which sat, with much the same membership, in 1820 and 1821, and indeed they went so far as to prepare draft Bills. Neither came to anything, largely, the committee members believed, because of opposition fomented by or on behalf of the threatened trusts. 'Your Committee,' they said in 1821, 'were . . . aware that any measure would meet with strong resistance which proposed in a greater or less degree to change the distribution of the large revenue now applied to maintain the Roads in question, and were not surprised to find that all the persons concerned in the present expenditure, were strongly indisposed to the change.'

There was another reason also. 'Your Committee are nevertheless persuaded that the grounds on which they recommended the measure, were in some degree misrepresented; and they beg leave therefore distinctly to disavow the wish imputed to them, of favouring or patronizing a particular individual, which (however highly they may think of his simple and scientific modes of road management) was not a part of their object.' McAdam, without much doubt, hoped to become surveyor to any consolidated authority that might be set up, but the Committee evidently thought it politic to placate those who were determined that he should not.[5]

Eventually it needed the power and influence of one of the McAdams' aristocratic patrons to get the matter settled, and in their favour. William Lowther (1787–1872), Viscount Lowther and after 1844 second Earl of Lonsdale, was a Tory magnate, immensely wealthy from the ownership of coal measures, at the centre of the world of unearned privilege which the McAdam family knew so well how to make use of in support of projects tending towards the establishment of a very different scheme of things. Lowther is recorded in the DNB as a good landlord, a successful racehorse owner (his Spaniel, a horse in spite of its name, won the 1831 Derby), a patron of Italian opera in London, a collector of porcelain. The DNB is patronizing about his abilities, attributing his position to his money and family influence, but his long career in politics brought him in 1841 as high as the Cabinet, as Postmaster-General under Peel, and as far as London roads are concerned he took official duties seriously and worked hard at them.[6]

Early in 1825 Lowther moved for the appointment of a Select Committee to investigate the management of turnpike trusts 'in the vicinity of the Metropolis'. It duly sat and reported that the roads had improved since Parliament's attention, in 1819, had been called 'to the expensive management and defective state of the Roads'. Having paid that back-handed compliment, the committee paid no more. Trust accounts, they said, were 'in a very confused state, and the Clerks of the Trusts utterly incapable of affording the information which the Committee required'. Income was larger than was needed to keep the roads in repair 'if those Funds had been skilfully applied and proper Materials obtained'. Debts were larger than they need be, and tolls higher. Borrowing, at 5 per cent, was too expensive, and if the trusts were consolidated all their revenues could be deposited with the Bank of England, thus getting rid of 'the many disjointed interests of the several Treasurers'. Two wealthy parishes in the West End – St George's and St Mary-le-Bone – were subsidized by turnpike trusts for repairs to Piccadilly and Oxford Street which they could well afford to pay for out of the rates. In the neighbourhood of the City there were four trusts – City Road, Old Street, Bethnal Green, Shoreditch – for $4\frac{1}{2}$ miles of road, 'thereby giving rise to numerous Gates and vexatious delays for collecting the Tolls, upon each of the several Roads; the amount of which appears far more than necessary for their maintenance'.[7]

The result was the Metropolis Roads Act which became law in the autumn of 1826. 'The Act,' said *The Times* on 10th November 1826, '. . . after enumerating about forty Acts of

Parliament, destroys them all at a blow. Its havoc among trustees is even more terrible. About 1,400 of them are disbanded with less ceremony than a useless regiment at the conclusion of a war; and all their important powers, duly registered on Turnpike Gates – all their local patronage – and all their trust dinners, eaten for the public good, are either to be abolished or engrossed by a peace establishment of forty-eight public men, who are appointed their successors'.

The Chairman of the Commissioners was Lord Lowther, who held office continuously until he died, in 1872. The 'forty-eight public men' included individuals as various as J W Croker (1780–1857), high Tory politician and writer; Joseph Hume (1777–1855), Scottish East India Company surgeon turned radical politician, who moved for the repeal of the Corn Laws in 1834; Sir Henry Parnell, already, as we shall shortly see, at odds with James McAdam; and a number of gentlemen such as George Byng of Wrotham Park near Barnet and Sampson Hanbury of Poles, Ware, who were trustees of roads outside London. Altogether the Commissioners of Metropolis Roads were a distinguished band, but can they have been harmonious?

As a body, in their early years, they were active and enterprising, and their Chairman was among the most active of all, attending meetings in the early part of 1827 at least once a week and frequently more often. They set to work at the British Coffee House in Cockspur Street on 31st July 1826 and formally took over from the trustees of the fourteen defunct trusts on 1st January 1827. By that time they had found themselves a house – in Whitehall Place – and appointed their officers, including James McAdam as General Surveyor. He was required to give security for £5,000, and his salary was set at £500 with £200 for travelling expenses. By the time he was formally appointed he had already made a report.[8]

Because London was so much larger and busier than any other city in the country the Metropolis Roads Commission differed both in name and nature from any other turnpike trust. Among those which members of the McAdam family served, the Bristol Trust came nearest, with 178 miles of road (more

Sir James McAdam (1785–1852) by an unknown artist.
Middlesex Area Magistrates' Court Committee

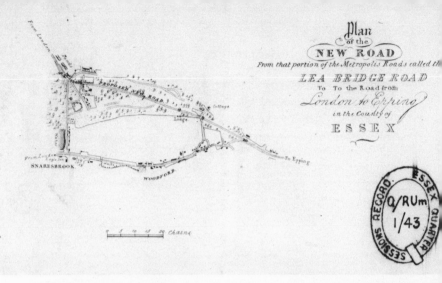

Plan, said to be by James McAdam, for the improvement scheme
described on pp. 170–1.

Essex Record Office

Apparently a satire, 1827, on McAdam's appointment as Surveyor to
the Metropolis Roads' Commission, which controlled the approaches
to the Great North Road and the Great West Road (see map, p. 159).

Tarmac Limited

than the mileage under the Commission), seven or eight sub-
surveyors (the Commission had nine), and substantial figures of
debt, income and expenditure. Even the Exeter and Bath trusts,
by comparison, were small, and most of the rest were tiny.

The Commissioners, when they first came into office, took
over bonded debt of £84,265, floating debt of £18,395 6s – a
figure which appalled them ('so inadequately provided for') –
ordinary annual income of about £78,000 and ordinary
expenditure some £10,000 less.[9] As well as being responsible
for all the main roads into and out of their area, they had
charge of all or part of many of the streets connecting with
them, including Old Street and City Road; Euston Road and
Marylebone Road; Oxford Street and Bayswater Road;
Knightsbridge and High Street, Kensington.

'A great public inconvenience', said the Commissioners in
their first published report, '. . . arises from numerous turnpike-
gates being stationed in the very streets of the metropolis.'[10]
Accordingly they attacked them vigorously, seeking to get the
upkeep of the roads transferred to the parish rates. Twenty-
seven gates, it was said in 1845, were got rid of in one day at
the start of the Commissioners' reign. In April 1827 there
were still forty-seven standing, as well as forty-four side bars
and seventeen weighing-engines, but in 1828 they got rid of
all the weighing-engines and by 1830 had removed gates from
streets 'on the Northern verge of the Metropolis' and handed
the upkeep to the parishes. They attacked another grievance
also: inequality and frequency of tolls. By an Act of 1830 they
equalized the tolls on the principle that for every seven miles
3d should be paid for every horse drawing a wheeled carriage
and 1½d for every horse not drawing. Their area was divided
into sixteen districts of roughly equal length, so that one toll
carried the payer right through.

The Commissioners' assault on their debts was equally
vigorous. About one-third of the floating debt was due in small
sums to small creditors, so the Commissioners borrowed
£10,000 from Overend, Gurney 'as a measure not merely of
justice, but of charity . . . for the purpose of satisfying the class

of demand smallest in individual amount but most urgent in respect to the actual necessities of the claimants'.[11] They proposed also to raise £100,000 from the Exchequer to enable them to concentrate the bonded debt in the hands of one large creditor rather than many smaller, and to provide them with £20,000 for new roads. The Exchequer discomfited them considerably by refusing the advance, but the Commissioners' resources were by no means exhausted. In August 1827 they borrowed £25,000 at 4 per cent, repayable over five years, from Overend, Gurney & Co. The loan was negotiated by Sampson Hanbury. With that in hand, they went on to get their £100,000 in 1830, also at 4 per cent, by way of £80,000 from Alliance Assurance and £20,000 from W H Cooper. At this period there seems to have been no shortage of willing lenders including, in 1827, Lewis Levy who offered £10,000 at 4 per cent against Overend, Gurney's 4½ per cent. The Commissioners turned him down because he was one of the lessees of their tolls 'and there are other considerations'. Perhaps they thought he was impertinent.[12]

The Commissioners' finances seem always to have been expertly managed. In their 25th Report, for 1851, they reported that their mortgage debt was totally extinguished, in spite of tolls which by then were nearly 25 per cent lower than at their highest figure, in 1837, and maintenance expenditure which was only about 17 per cent lower.

The Commissioners' General Surveyor, aged forty-one at the time of his appointment, entered into his duties as energetically as the Commissioners themselves, producing within three months a detailed report on the 131 miles of road taken over, founded on a thorough examination of the construction of the road surface by 'pitting' it every half-mile. The trusts, thoroughly alarmed by the threat of consolidation hanging over them since 1819, and no doubt also by the tone of the report of the Select Committee of 1825, had bestirred themselves and they handed over comparatively little for McAdam to complain of. As well as looking to the general state of their existing roads, two or three trusts had gone to Parliament to get powers to make major improvements, and these were written into the Metropolis Roads Act, including a new line of road from Portland Place across Camden Town to Holloway Road and 'a very extensive alteration in the line of the Harrow Road'.[13]

McAdam's heaviest criticism was directed at the Kensington and Fulham roads. Within about two miles from Kensington Barracks, travelling west, he found 'no less than eleven alterations in the mode of forming a road', including side-to-side paving, various combinations of paving with gravel and flint, and unpaved surfaces of Kentish rag, gravel and flint. Stretches where the paving was in the centre produced very exciting driving. 'By this mode of forming a highway,' said McAdam, 'two very narrow roads are left for all fast travelling carriages, which, on passing or overtaking each other, are obliged to pull up on the pavement on the centre, which never fails to be several inches higher than the gravel road at the sides.'

To remedy this state of affairs, and as a general prescription for the health of the Metropolis Roads, McAdam recommended three standard materials for repair and upkeep, depending on the distance from central London. For the 'first portions of the principal lines of thoroughfare' he wanted Guernsey granite or whinstone. 'Although the first cost of these excellent materials might be great,' he said, 'I am confident that their durability, the saving of labour, coating, scraping and cleansing, would in the end be productive of saving of expense, and would give to the public the best possible description of road.' For 'the next and middle portion of the road' he recommended flint, and 'at the greater distances from London gravel of a much better quality, and at much reduced expense, may be obtained, and by attention in procuring, cleansing and preparing it, a good road may be upheld.'

He turned next to management and the price of labour. All the employees of the old trusts were dismissed, and the roads were divided into nine districts, each to be under a resident sub-surveyor at not more than £150 a year. Six sub-surveyors were appointed in October 1826 leaving three places for surveyors of existing trusts, if they were found good enough.

Variations in the rates of labourers' wages scandalized McAdam and he rapidly evened them out. 'At Stamford Hill 11s per week; on the adjoining road, Old-Street, 18s per week; at Hackney, 15s; Harrow, 12s; at Kensington, 17s 6d; independently of a numerous class of foremen at 20s and 25s per week. There was much difficulty, and considerable discretion necessary, in arriving at a fairer and more uniform distribution of this great and important portion of the expenditure on the roads, and I am happy to be able to report [in April 1827] that 14s per week is now the maximum rate . . . It was highly proper and necessary that the payment for work on the Metropolis Roads should bear a fair and reasonable proportion to the general labour of the adjacent parishes.'[14]

In his work for the Metropolis Roads Commissioners, described from year to year in his printed reports, McAdam showed two attributes which no doubt contributed to his success as a surveyor in London and elsewhere. On the one hand no detail – removal of 'dangerous and inefficient paved water channels' across Highgate Hill, 'a laudable spirit in favour of improvement' among parish officers at Fulham, 'a very inefficient drain' giving offence 'with reason' to the inhabitants of Little Chelsea – is too small or too localized for his attention and comment. On the other hand he shows a firm grasp of the Commissioners' long-term policy, in which the interest of London as a whole – 'the Metropolis' – was given precedence over the interest of separate localities.

For the Commissioners, this was by no means a highroad to popularity. As anyone who has lived in London knows, Camden Town might as well be in Africa as far as the inhabitants of Chelsea are concerned, and the same might be said of any two localities separated by a few miles of streets. London even today is very much a loosely strung network of villages, each with a narrow view of its own advantage, and it was very much more so in the early 19th century. Hence, no doubt, some of the opposition to consolidating the fourteen trusts. People obliged to pay tolls, which they disliked anyway, in the district where they lived were somewhat mollified if they knew the produce of the tolls was locally spent, but they were not at all pleased when they discovered that the Commissioners were prepared to spend the money, if they thought proper, on improvements far away.[15]

In their early years, however, the Commissioners were probably no more disliked than the general run of turnpike trustees : perhaps less so. Macadamizing was at the height of its vogue, and before the Commission was set up it had been applied to Whitehall, Regent Street, Jermyn Street, Bond Street and the fashionable shopping area in Clifford Street, Burlington Street and Cork Street behind Piccadilly. It soon spread to part of Piccadilly itself, to Pall Mall, to Oxford Street, and it is clear from James McAdam's own report of 1827 that several much denigrated trusts had begun to follow the true doctrine in less glamorous areas.[16] McAdam himself propagated it with vigour, seeking throughout the Commissioners' area to get rid of paved carriageways – the last was reported to have gone in 1838 – and to substitute surfaces of hard stone, prepared and laid according to his father's System.

At the height of his activity, in the mid-thirties, James McAdam was laying road material at a rate of about 100,000 cubic yards a year, buying it on contracts put out to competition 'for the double object of preventing combination, and procuring an ample supply on reasonable terms' – for which he could rely on the Commissioners' strength as the biggest buyers in the market. The same system was applied for the same reason to contracts for the execution of work, for tools, and for supplies generally. Stone of several different kinds was bought from widely separated sources. Discussing his plans in 1828, McAdam said :

'. . . an arrangement is hoped to be made for the supply of flints by the Wandsworth Rail-road,* and by the River Thames, from various places in the county of Kent; by the Grand Junction Canal, from Bucks and Herts; whilst the

---

* Presumably the Surrey Iron Railway, opened from Wandsworth Wharf to Croydon on 26th July 1803 (Hamilton Ellis, *The Pictorial Encyclopedia of Railways*, Hamlyn 1968, 12).

supply by the River Lea, from the Rye Common, has proceeded with increased activity and success. Contracts have been entered into for the supply of granite from the Island of Guernsey, and Mount Sorrel, in Leicestershire : and, after much difficulty in arranging the details, a large supply of a valuable and new material for the service of the roads has been opened from Ightham in Kent.'[17]

The Commissioners, in 1828, took a wharf on the Regent's Canal at Battle Bridge (King's Cross) which by 1850 had disappeared under the line of the Great Northern Railway, being replaced by a new wharf in Old St Pancras Road.[18]

Besides improving old roads, the Commissioners built new ones, basing all but one of their most ambitious schemes on plans prepared by their despised predecessors, the Trusts. One of these schemes, for the radical reconstruction of the Harrow Road, seems to have been quietly dropped despite a strong recommendation from McAdam. Another, to provide a connecting link between Shepherd's Bush and the Great West Road at Turnham Green – about a mile 'and three-quarters – produced what is now Goldhawk Road. A third project, for 'the Camden Town new line of road', running from Portland Place along the edge of Regent's Park by Albany Street and then by Parkway and Camden Road to Holloway Road, was extended by nearly three miles. With powers granted by their Act of 1830 the Commissioners launched the road north-westward, aiming at 'a clump of trees called the Seven Sisters near Tottenham' on the Old North Road, and by so doing they opened up 'the populous parts of Hertfordshire' to travellers from the Western districts of London, as well as shortening the distance between Westminster and Cambridge 'and that line of country'.[19]

The same Act permitted the Commissioners to launch a scheme of their own, for 1¾ miles of road from Whipp's Cross through Epping Forest to the Castle Inn near Woodford in Essex. There it was intended to connect with a somewhat longer stretch of road, from Epping, which the Epping and Ongar Trust (surveyor, James McAdam) was seeking powers

to build.[20] The Commissioners' road was opened in 1831 and
the Epping and Ongar road in 1834. The combined effect of the
two projects was to improve communications between London
and Bury St Edmunds and Norwich, and of their own road,
which was estimated to cost £1,200, the Commissioners
observed : 'So great an improvement in the neighbourhood of
the Metropolis was perhaps never before effected for so small
a sum.'[21]

By 1833 the Commissioners' impetus towards new construction was running down. In their report for that year they
announced no new lines of road, 'being desirous for the present,
rather to await the experience of the working of the important
and expensive Roads that have been made, and to see their way
in the effect . . . upon their general finances'. The toll revenue
in that year was slightly down, which no doubt influenced them,
but two years later, when it was on a strongly rising trend,
they were even more definite about a stop to new enterprises.
'The various matters', they said in their report for 1835, 'in
contemplation of which the Commission was formed, having
been brought into regular train, the Commissioners have not
much to state beyond a reference to the Report of their General
Surveyor', with whose work they expressed 'continued satisfaction'.

The General Surveyor's report shows that he was still very
busy with upkeep and improvements, which are described in
detail. His work for the Commissioners, nevertheless, was
approaching its grand climacteric. In 1835 his department
used 106,909 cubic yards of material : never before nor ever
again so much. Two years later the Commission's expenditure
on upkeep reached a peak – nearly £74,000 – to which it never
afterwards returned.

The Dispute with Telford: the Holyhead Road at St Albans and Whetstone:

Along with the Commissioners, their General Surveyor reached
the zenith of his ascendancy, in public affairs generally as well

as in theirs, during the mid-1830s. Almost as soon as he was appointed, however, he had to face a determined attack on his professional competence. It was launched in Parliament in 1828 and it was directed by one Commissioner, Sir Henry Parnell, and supported in evidence by another, George Byng.

The attack came from the direction of another board of commissioners, the Commissioners for the Holyhead Road, who had the responsibility of overseeing the affairs of all the turnpike trusts along that road, and very wide powers for doing so, which they added to from time to time, when major works were required, by getting fresh Acts. James McAdam came into collision with this formidable body because he was Surveyor to two small trusts which had the misfortune to lie directly on the Holyhead Road immediately outside London. One was the Whetstone Trust, responsible for $8\frac{1}{2}$ miles of road between Highgate and Barnet (the present A1000) and the other was the St Albans (Pondyards and Barnet) Trust which maintained about eleven miles of road following directly on from Whetstone territory, along a line now much modified but roughly represented by A1081 and A6(T) through St Albans to a point (the Pondyards) outside the town on the North-West side.

One of the most active of the Holyhead Road Commissioners was Sir Henry Parnell, created Baron Congleton in 1841. He was Anglo-Irish, great-uncle to Charles Stewart Parnell, and sat for Queen's County from 1806 to 1832 and for Dundee from 1833 to 1841. He was a very industrious politician of liberal views, interested especially in Irish affairs, Catholic emancipation, finance and political economy. He was instrumental, in 1815, in getting money voted for the improvement of the Holyhead Road and in setting up the Commission. In his dealings with the St Albans Trust, as they are recorded in the trustees' minutes – admittedly a prejudiced document – the impression comes through of a clear-sighted, high-handed administrator, convinced of the rightness of his views, impatient of anyone else's, and totally lacking in tact, which perhaps he might have regarded as an admission of weakness.

'As the St Albans Trust have very wisely placed the care of

their Road under Mr McAdam', said the Select Committee on the Road from London, by Coventry, to Holyhead in 1820, 'the other Trusts will be able to see how much more economical and efficient his method is than that which now prevails, and how very easy it will be to introduce it generally along the whole line of the Holyhead Road.'[22] At Whetstone James McAdam was 'recommended by Sir Henry Parnell, and many other gentlemen as a competent person'. Sir Henry himself said that when he read J L McAdam's 'very good directions for the management of gravel materials' he was induced 'to think and expect that if Mr James McAdam was appointed surveyor, there would be a great improvement in the management of the road'.[23] James McAdam therefore took up his duties both at St Albans and at Whetstone in an atmosphere of goodwill all round.

Thomas Telford the Holyhead Road Commission's engineer, vastly more eminent in his profession than James McAdam, at that time, was in his, had been at work for the St Albans trustees before McAdam arrived. One of the first things McAdam had to do in his new appointment was to remove fences in order to open a short stretch of road, on Lord Hardwicke's land, over Ridge Hill about five or six miles from St Albans on the London side. Telford had advised the trustees to make the road in a short cutting, in order to ease the gradients on the hill.

In making this agreement the St Albans trustees acted as an independent body of similar legal standing to the Holyhead Road Commissioners : certainly in no way subordinate to them, a state of affairs which neither Sir Henry Parnell nor Telford intended should last much longer. Telford, on the instructions of the Treasury, surveyed between 1818 and 1820 the whole line of road from London to North Wales, and in his report he made proposals affecting trusts all along the route, including the trusts at Whetstone and St Albans. Many were expensive proposals which would be bound to add considerably to the debt of the trusts concerned, not at all to the advantage of existing bondholders. But the Commissioners went ahead and sought parliamentary powers to get Telford's proposals carried out, if necessary over the heads of the trustees.

Trustees, certainly at St Albans and almost certainly elsewhere, armed with statutory powers themselves and no whit inferior, in their own estimation, to the Commissioners of the Holyhead Roads, deeply resented this assault on their independence, especially in view of their obligations to their bondholders. They resented it as much as the trustees of the Metropolitan Roads, a few years later, resented the proposal to consolidate the London trusts. It is ironic that James McAdam, that enthusiastic centralizer, should have found himself, as Surveyor to the St Albans Trust, the victim of a situation exactly parallel to that which, as Surveyor to the Commissioners of Metropolis Roads, he was a little later all eagerness to exploit.

The quarrel at St Albans boiled up in the early 1820s over two proposals for which the Commissioners sought, and eventually obtained, parliamentary powers by which they could, and eventually did, override the trustees. Their first scheme was for a new route through the streets of St Albans requiring the destruction of property and hence expensive compensation on top of the costs of construction. Secondly, they intended a completely new line of road, about $2\frac{1}{2}$ miles long (the present A1081), which would connect South Mimms with Barnet straight instead of by the curving course of the road by Kitt's End and Dancers Hill which already existed (and still exists).[24]

These proposals would both improve the roads of the St Albans Trust, but they offended the trustees for two reasons. First, they would require large borrowing from the Exchequer, up to many times the existing debt of this small trust, and increased tolls at certain gates, which would be locally unpopular and, in the trustees' view, unjustifiable on purely local grounds. Secondly, the borrowed money would be handed over to the Commissioners, and neither the trustees nor their surveyor would have any control over the work it would pay for, in spite of the fact that after Mr Telford and his assistants had built the new roads they would be handed over to the trust and the trustees would have to cope with the financial consequences. They strongly suspected – rightly, as matters turned out – that the final cost would be much greater than the original estimates, which was another reason why they disliked the whole affair.

The manner in which the Commissioners' proposals were presented was no more calculated to win over the trustees' hearts and minds than the matter of the proposals themselves, for it was made pretty plain that Sir Henry and his associates regarded them as minor local busybodies to be got out of the way as soon as possible. In June 1823, for instance, Telford's assistant Easton twice failed to produce a plan of the improved route through St Albans when the trustees had expected him to do so and then, to their intense annoyance, produced one on which the route was shown in pencil. Five years later Sir Henry Parnell, having asked for the date of a meeting to be changed to suit his convenience, then failed to appear, writing to the trustees' clerk to say that he could not leave London that day.[25]

What probably gave the greatest offence, however, was repeated criticism, by Telford and Sir Henry, of the upkeep of the St Albans roads, and Sir Henry's repeated suggestions that the Commissioners, as well as managing those roads which they were altering, should also take into their hands, for a period, the rest of the roads of the trust so that they could put them in order. James McAdam, whose professional competence was thus under attack, eventually became so exasperated that in June 1828 he 'voluntarily stated to the Meeting [of trustees] that as so much of the line of Road within this Trust being under the management of the Holyhead Road Commissioners, he could not but consider the Trustees paid too high an allowance for the surveying of the Road . . . and he stated that he should be satisfied with a Salary of Fifty pounds per Annum as Head Surveyor . . . And the Trustees present accepted Mr McAdam's liberal offer and concur[red] in his suggestion.'[26]

The Commissioners' parliamentary proceedings were not

rapid and the St Albans trustees fought a determined delaying action. Nevertheless in the end the Commissioners had their way and both their major schemes were put through. The road through St Albans was handed over to the trustees in June 1829 and the new line between South Mimms and Barnet, after several postponements, in July 1833.[27]

James McAdam, in several reports, was quite as unflattering about Telford's methods as Telford and the the Commissioners had been about his. Before he formally took charge of the road through St Albans he and his father surveyed it, no doubt extremely carefully. With relish, probably, he put in a damning report. The road had been represented by Sir Henry Parnell as being 'in perfect condition'. The two McAdams pitted it every quarter mile at the centre and at each side and found that for part of its length it was covered with 'an inch coat of Nuneaton stone very badly broken'. Below that they found 'unprepared Gravel with a very large proportion of Loam Sand Earth Hoggin and large unbroken Flints', and for some distance the road consisted of nothing else. 'This Road,' wrote McAdam triumphantly, 'requires to be picked up and the Materials and surface prepared to receive . . . a coat of clean well broken Flints carefully raked and attended to until the surface shall become smooth and solid.' A few months later he asked for a rise 'in consequence of the additional Duties which have fallen upon him from the various improvements made in the Road during the last and present year and of his now having the care and management of the New Road from St Albans to the Pondyards'. The trustees brought his salary up from £50 to £75.[28]

When the Mimms–Barnet stretch was handed over McAdam gave it the same treatment – pitting every quarter mile at the centre and the sides – with equally horrifying – or gratifying – results. 'It appears,' he reported, 'that the average depth is about 8 inches, and . . . the Materials consist of unsorted Gravel not sufficiently sifted, being a description of Material and strength of Road greatly inferior to that stated as desirable and necessary in the various Reports of the parliamentary

Commissioners, and which the Trustees had a right to look for on this three Miles of Road made by the parliamentary Commissioners, with the funds of the Trustees at an expense of £18,000.'

A shrewd blow, but there was more to come:

'In presenting this Report to the Trustees, I consider it my duty to inform them that the day Labourers employed on this Road under the Parliamentary Commissioners were charged to the Trustees at the rate of 16s. per week wages.'

Surprised, he said, at a charge so unusually high (the going rate in the neighbourhood was twelve shillings), McAdam investigated, 'and found that these Day Labourers were frequently unpaid for two and three Months together which compelled the Labourers to resort to the Tommy Credit Shops for the subsistence of themselves and Families where in consequence of the length of Credit, and the risk run, they were charged much higher for the necessaries of life, and being unable to keep a correct account of every Article received for so long a period were compelled to pay such demands as were made upon them . . . [when] they received their Wages.' McAdam took them into the service of the trustees at 12s a week regularly paid, and they expressed themselves 'much better satisfied'.[29] The first Truck Act had been passed in 1831, but it does not seem to have come to the notice of the Commissioners' contractors.

The Commissioners' activities, as the trustees had all along feared, were financially disastrous. They burdened the Trust with excessive tolls – 'the present Toll at South Mimms', said James McAdam in 1833, 'of three shillings and sixpence on each Stage Coach at each time of passing is so loudly and so justly complained of' – and a load of debt monstrously swollen.

In 1837 Francis Searancke, the trustee who acted as auditor, reported the financial state of the Trust. The 'Kingsbury improvement' (the road through St Albans), he said, had cost, including interest, £21,807 15s 2d against the Commissioners' estimate of 'about £10,000'; the Mimms–Barnet road,

£30,435 6s – 'this alteration we were led to expect would amount to about £17,000'. Taking account of considerable sums already paid off, the trust's total debt, Searancke calculated, stood as follows:

Kingsbury	£8,995	14	11
Mimms	20,317	13	7
Mortgage Debt	6,572	16	2
	£35,886	4	8

The debt to the Exchequer, assumed unwillingly at the behest of the Parliamentary Commissioners, was thus about four and a half times greater than the mortgage debt assumed by the trustees on their own responsibility. The Trust's total income Searancke gave as £6,961 'which will allow an annual payment of about £3,000 in liquidation of our Debt to Government.' Clearly there was not going to be much left for upkeep, though without the Commissioners' imposition the trust would have been comfortably placed.

'The Trustees present', says the report of the meeting at which Searancke's figures were considered, '. . . view with alarm the enormous Debt which has been saddled upon them and without their Controul by the Parliamentary Commissioners of the Holyhead Road, and the vast accumulation of Interest going on in respect of these Sums.'[30] No wonder, and the trustees got very little help from the Treasury. As late as 1852 they memorialized the Home Secretary 'praying him to take into consideration the burden thrown upon the Trade of the Borough of St Alban and the Agricultural Interest in the Neighbourhood by the imposition of a Toll levied at the end of the Town for the purpose of paying the Interest of money advanced by the Lords of the Treasury for the Improvement of the Road from London to Holyhead which does not collect sufficient to pay the original Interest and therefore has become a hopeless burden to the Town and Neighbourhood.'[31]

This cry of distress goes close to the heart of the matter. Unquestionably it was unfair that a turnpike trust of modest size and resources, set up to serve its immediate neighbourhood, should be burdened with the expense of improving its roads for the benefit of Irish MPs and other long-distance travellers. Moreover, Sir Henry Parnell and his colleagues seem to have thrust the trust into debt with no regard at all for the local consequences. But as long as 'great roads', later known as 'main roads' or 'trunk roads', were regarded purely as the financial responsibility of the localities they passed through, either the unfairness would persist or the 'great roads' would be in danger of neglect. This was the essential defect of the turnpike system considered from a national point of view: a defect which James McAdam and his father were well aware of.

*   *   *

From the fifties we must turn back to the twenties. While Sir Henry and Telford were attacking McAdam for his alleged failings as surveyor to the St Albans trust they found an opportunity for another attack, potentially more dangerous, based on his activities as surveyor to the trust at Whetstone. The attack arose, like the quarrel at St Albans, from proposals by the Holyhead Road Commissioners for new and expensive works, and it led to a parliamentary enquiry in the summer of 1828. The Report and Minutes of Evidence of the enquiry, indeed, are the only source of evidence for the conduct of the attack, because the minutes of the Whetstone trustees have disappeared.

The enquiry, by a Select Committee of the Commons, was set going by Sir Henry Parnell after he had received a petition from a number of London coach proprietors complaining of the state of the roads under the care of the Whetstone and St Albans Trusts, which they said made very heavy going for their horses. This was useful ammunition for the Commissioners, who had quarrelled with the Whetstone trustees, as they had with those at St Albans, and were determined to make out a case for taking over both trusts' roads. That was at the heart of the enquiry, and it inevitably meant an attack on the professional competence of James McAdam, surveyor to both trusts. The attack was comprehensive, but one matter stood out : the management of a project for easing the slope of Barnet Hill, by which coaches from London approached the town of Barnet.

This project was put to the Whetstone trustees in 1823 by the Holyhead Road Commissioners, who in their usual way suggested that money should be borrowed by the trust so that the commissioners might do the work. 'The Trustees, in reply to this proposal, stated that as they had in their service Mr James M'Adam, as a surveyor, they would prefer undertaking the improvements themselves . . . according to the lines of inclination proposed by Mr Telford.'[32] It was from this decision that the trouble began.

Telford's plan involved lowering the street level in Barnet by six feet, which alarmed the householders affected. They were constituents of George Byng, one of the Whetstone trustees, who suggested to McAdam raising the embankment at the foot of the hill instead of lowering the road at the top, which McAdam agreed to do. Unfortunately he did not know as much as he should have done about embankments. His was built of yellow clay, it slipped, and he admitted to the Select Committee that 'he was not aware . . . that clay would not stand at a less slope than 1 in 4'.[33]

That error and the cost of the works, far in excess of the Commissioners' estimate, laid McAdam open to charges of incompetence both from the Commissioners and from trustees. George Byng, who was also a trustee at St Albans and a Commissioner of Metropolis Roads, said, 'We were bit by Mr McAdam', which implies something worse even than incompetence. After a good deal of dispute among the trustees at Whetstone, and a reduction in salary, McAdam avoided dismissal, but in this case it is difficult not to feel that his critics were justified.

Against Sir Henry Parnell McAdam in the early 1830s had an ally at the summit : Charles Gordon-Lennox, 5th Duke of Richmond (1791–1860), a great racing man in his private

Brentford High Street, 1819. 'At present, the Town of Brentford is . . . so narrow . . . as to render a quick passage through it a matter of no inconsiderable danger.' Sir James McAdam's Report to the Metropolis Roads' Commission, 1837. *Museum of London*

This coach has passed Highgate Archway and is making for the Whetstone Trust district (surveyor, James McAdam).

Fotomas Index

Hyde Park Corner about 1829–1834. The road works are presumably under the direction of James McAdam, as Surveyor to the Metropolis Roads' Commission (p. 164). · *Museum of London*

Tyburn Turnpike Gate, near the present Marble Arch, about 1820. A gate on one of the roads taken over in 1826 by the Metropolis Roads' Commission.

Museum of London

capacity and in public life Postmaster-General from 1830 to 1834 in the Government which passed the first Reform Act.[34] He had gone into politics after five years' service under Wellington in Spain and at Waterloo : service far from purely ornamental, for as an infantry captain he was badly wounded at Orthes in 1813. He was a Tory, but he joined Grey's ministry for recondite reasons of party tactics, and he evidently wielded considerable influence, for he was taken into the Cabinet : not usual for the Postmaster-General.

On 4th April 1831 Sir Henry Parnell came into the Government as Secretary at War, not in the Cabinet. His proceedings thereafter convey the same impression of opinionated obstinacy as his dealings with the St Albans and Whetstone trusts. He infuriated the Duke of Richmond by unauthorized negotiations with the French post office and by encouraging Joseph Hume to attack the British post office. He was nearly dismissed.

In January 1832 he courted dismissal again by refusing to support the Government in a division, and this time he was thrown out, though he came back into office again, as Paymaster-General, in 1835 and remained in that post as long as he lived. He became Lord Congleton in August 1841, but had little joy of it, for he fell sick and on 8th June 1842 hanged himself in his dressing-room.

Parnell's downfall can hardly have been displeasing to McAdam. The Duke of Richmond relied on him a good deal for advice on turnpike roads and he, for his part, sought the Duke's support, and generally got it, for projects that interested him both while the Duke was in office and after he resigned. Letters passed between them fairly frequently, and a number are preserved in the Goodwood archives in the West Sussex Record Office. Over the years their tone passes from rigid formality – 'I would esteem it a great favor if Your Grace would have the goodness to see me for a few minutes' – to something a little more relaxed – 'I will take my chance at half past one tomorrow', and in the last years of James McAdam's life one or two of the letters come as close to familiarity as was likely between a commoner and an early

Victorian duke : 'I most sincerely hope this will find Your Grace enjoying good health altho' I fear judging by myself feeling the effects of getting older.'[35] McAdam then was 64, the Duke 57, and they had known each other certainly for the greater part of twenty years, perhaps much longer. As early as 1834 McAdam felt sure enough of himself to send 'a few Charmontels from Guernsey by the Chichester coach to Goodwood' – where the Duke had his seat – 'they came with a cargo of granite from the Island last week', and the Duke thanked him 'for some of the finest Pears I ever tasted'. In 1836 he sent, on behalf of the Metropolis Roads Commission, a rather different present – 'one of our stone breaking Stands with Iron blocks fixed in it, and also a ring with handle to hold the Stone Steady and one of the Hammers we use'. He sent it off 'by the waggon from Queen's Head Boro' and evidently the Duke was going to have to pay for that, for he added : 'Your Grace may think it a *dear Lot* at the *Carriage Cost*, which is all the expense as they are presented by a very rich Trust.'[36]

The present of the stone-breaking stand was made after the Duke had left office, presumably because in his capacity as a country gentleman he took an interest in turnpike roads and their upkeep. Even while he was in office, McAdam took care to see that his private interest in these matters was not neglected. 'Knowing the Interest Your Grace takes in the Newmarket Line,' he wrote in 1833, 'I ventured to lay before you, the plan of a little Improvement I am anxious to effect.'[37]

That was McAdam writing as surveyor to the Newmarket Heath Trust. At much the same time, as surveyor to the Epsom Trust, he found himself faced with a complaint that King William IV, in travelling from Windsor to Brighton, had found a stretch of road between Ewell and 'Ryegate' (roughly A240) 'very steep, 1 in 12, narrow, with several dangerous turns'. McAdam made a survey and told the Duke that he was 'happy in being able to state, that an excellent line of new Road may be made . . . the steepest part . . . would be only 1 in 22½ which is an easy trotting slope – Could I obtain His Majesty's approbation of this Improvement I am confident I could prevail on the Trustees of the Epsom Road whose General Surveyor I am to carry it immediately into effect, and which would be the means of giving employment to the poor in the neighbourhood at present so much in want of work.'

The Duke showed the plan to the King. The King 'was pleased to express his decided approbation' and asked to see McAdam, 'being anxious to speak to you upon the Plan'. 'I have no doubt,' McAdam replied, 'that the "decided Approbation" will enable me to overcome all difficulties at Epsom.' It didn't. About a month after McAdam's outbreak of optimism, the Epsom trustees met. They were told that Mr Buckle, Lord of the Manor and also owner of most of the enclosed lands over which the new road would run, had refused his consent. So much for the royal approbation, decided or not.[38]

Improvements at Newmarket and on the road to Brighton, ducal interest and royal complaints notwithstanding, were local matters. McAdam's public service on a national scale, apart from his widespread influence as surveyor to so many trusts, was in the background of politics: in the world of draft Bills, Select Committees, Royal Commissions; of discussions with officials, with peers, with MPs; of memoranda for the Duke and any other great men he wished to influence; of the management of parliamentary tactics. In all these manoeuvrings he was animated by two main aims, pursued alike by himself and his father: to get turnpike trusts under central control, and to get them 'consolidated' into groups.

McAdam had great hopes of the Turnpike Returns Act of 1833. 'I am quite of the opinion', he wrote to the Duke while it was still going through the Commons, 'that the Returns Act may be made the *Nucleus* of a *Controul* Quietly without exciting alarm or even opposition.' The trusts already had to make annual returns of their finances to the Secretary of State under the Turnpike Act of 1822, but the new Act required greater detail – including the names of surveyors – which was a step in McAdam's desired direction. He hoped to see 'controul' by financial means. First, the Act would give the Secretary of

State 'or those he may appoint' 'ample power . . . to ascertain which Trusts are good and safe security for their Mortgage Debts and the regular payment of Interest and Sinking fund.' Secondly, if the £4m of turnpike debt of good security, most of it at 5 per cent, were to be paid off in Exchequer Bills at 2½ per cent, 'the Government would derive a clear Profit of Eighty to One hundred Thousand Pounds per Annum without risk without expense and without trouble – whilst the most Salutary Controul would quietly be placed over the Trusts by compelling them to Pay off their debts and economise their funds.' This note is marked by the Duke : 'send a copy of this in my letter to Lord Althorpe without date or signature.'[39]

The turnpike returns became very important to McAdam. 'I have at last,' he wrote to the Duke in 1836, 'but not without great difficulty and a very extended correspondence, brought to a close the abstract of the Turnpike Trust Accounts for 1834.' He did the same thing every year for the rest of his life, with results published in the annual Abstract of Statements of Income and Expenditure of Turnpike Trusts. Immediately McAdam was dead, the information in the abstracts became sparser.

With the letter announcing the completion of the abstract for 1834 he included, 'for Your Grace's *private* information', a general statement for the same year 'Supposing, the Turnpike Roads in England and Wales were one Consolidated Trust.' It showed 1,039 trusts in England, 69 in Wales, with total debt of £8,453,391 6s 1d, total income (exclusive of 'balance in Treasurer's Hands brought forward') of £1,753,544 9s 11d and total expenditure (including 'balance due to the Treasurer brought forward') of £1,828,730 13s 1d.[40] One reason why turnpike trusts took up so much parliamentary time is immediately apparent.

Towards the end of 1833 McAdam and 'Mr Lyon' were 'diligently at work', in McAdam's words, on three bills which would complement and extend the effects of the Turnpike Returns Act. One was intended 'to add to the Metropolis Commission all the Turnpike Roads leading from the Metropolis not included in their present Act'. The second was to consolidate the eleven General Turnpike Acts, taking account of recommendations of a Select Committee of the House of Lords which sat in 1833 – 'truly a work of labour and time, as every Clause requires to be taken out separately to consider its effects on the others'. There were several points on which McAdam and Lyon were 'very desirous to have the benefit' of the Duke's opinion. Finally, they were working on a bill 'to Constitute a Board of Controul'.[41] On this they wanted to consult George Lamb (1784–1834), brother of Lord Melbourne, who was Under-Secretary at the Home Office, but the poor man upset a kettle of boiling water on his leg and died of the effects.

Decline: the Shadow of the Railways:

None of these measures passed into law, unless the bill to consolidate the eleven General Turnpike Acts was an embryonic form of the General Highway Act of 1835, which abolished statute labour and the composition money in lieu of it. That was not a provision which pleased Sir James. In 1839 he said that turnpike trusts had suffered very severely from it.[42] Although in the late thirties and forties he was still working assiduously in support of policies he favoured, none of his projects seems to have had any success. The Duke was out of office, the turnpike age was slipping fast away, the railways were spreading, and when Parliament looked at the roads, as it frequently did, it was usually with the object of easing the passage of turnpike trusts, solvent if possible, into oblivion. McAdam, with his schemes for central administration and a Board of Control, must have seemed to his younger contemporaries quaintly old-fashioned.

He was by this time well-off, evidently sociable and convivial at his house in St John's Wood, and there is no reason to think that he allowed the decline of the roads to blight his life entirely. In the autumn of 1835 the Metropolitan Commissioners asked him to inspect granite quarries in the Channel Islands, which gave him the opportunity, once again, to send Charmontel pears from Guernsey to the Duke of Richmond. 'Lord Lowther', he told the Duke, 'was so good as [to] offer to take me via Calais we had a delighful Tour through Normandy where I found the roads much improved tho' still much to learn.' 'We gave them,' he said, 'many a lesson of *actual* work and got Scrapers made a *Tool* and *operation* they never heard of.' Rather as an after-thought he added: 'we found Louis Phillpe [*sic*] at his Chateau at Eu'.[43]

The tour was obviously a success, and in July 1836 McAdam was intending 'another run in France in a few days' and was asking the Duke for 'the favour of Your Grace to oblige me with a letter, to Lord Granville,' – ambassador at Paris 1824–1841 – 'or to the French Authorities or any person your Grace may please.'[44] James was evidently as keen to spread the gospel according to McAdam among the foreigners as among people at home, but in spite of his eminence these two letters are the only positive evidence of his having actually gone abroad to do so.

While James McAdam was at his busiest, his knighthood came. 'I am happy to inform you', the Duke wrote to McAdam on 31st January 1834, 'that Lord Grey has consented to recommend you to His Majesty for the Honor of Knighthood.' On 20th March he told him 'you must attend at the Levee next Wednesday [26th March 1834] . . . you are at full liberty to inform your friends.'[45] The Dictionary of National Biography, in its extremely inaccurate entry about John Loudon McAdam, says that James accepted a knighthood which his father declined, and Mrs Pember says that John Loudon declined it 'on account of age and infirmity'.[46] The story may be true, but it seems more likely that it is the result of James being overshadowed in the view of posterity by his father or, perhaps more accurately, being overshadowed by an advancing railway locomotive.

The coming of the railways, along with the growth of London and the increase in numbers of public carriages, especially omnibuses, crops up again and again in the reports of the Metropolis Commissioners and the Surveyor from the mid-

thirties onward. The Commissioners, though rightly appre-
hensive about the effect of railways on their income from roads,
were far-sighted enough not to obstruct the building of them
unnecessarily. They considered it 'their first duty', they said in
1836, 'to protect the public traveller from inconvenience and
danger; but after the accomplishment of that object to give
facility to such operations'.

There were even fringe benefits. The railway companies
would sometimes improve roads, at their own expense, for
their construction traffic, as the Great Western were reported
to be doing at Hanwell in 1837 and the London & Birmingham
at Wembley. At Paddington in 1838 the Great Western were
building new parish roads, which enabled the Commissioners
to take the popular step of moving a toll-gate further out,
giving relief on the London side.

On the whole, though, in London as elsewhere the railways
presented a mortal threat. Roads near their construction sites
were badly worn, as they similarly were at Bath (p. 154), by
the heavy traffic required for building them. Worse, because
the effect was permanent – and very rapid – rather than
temporary, they struck at what McAdam called 'the long
thoroughfare': the stage-coaches, post-chaises and carriers'
wagons which provided so much of the turnpikes' staple
revenue. When the tolls were put up for auction in 1838, 'no
bidders would offer for several of them at nearly the amount
which they had been let for in former years. It appeared to be
the opinion of those who have usually bid for the Tolls, that the
opening of some Railways to the northward and westward of
London . . . would materially interfere with the accustomed
receipt of Tolls on the ordinary Roads'.

'Our worthy friends the Takers', as McAdam once described
Lewis Levy, Elisha Ambler and the other toll farmers,[47] were
quite right, as a glance at the figures of income for the Metropolis Trust will show. In 1842 the tolls revived, which 'induced
the Commissioners to entertain a hope that the transfer of
traffic from the Turnpike Roads to the Railways had now
arrived at its full extent', but they were over-optimistic. In

1846, the year of the Railway Mania, they were greatly exercised, in spite of most tolls being let at advanced prices, by 'the present extraordinary extent to which railroad making is threatened to be carried in the vicinity of the Metropolis'.

As the railways were building, London was spreading, not least along the roads and streets of the Metropolis Trust. McAdam's report for 1842, for instance, mentions 'extensive building speculations at Shepherd's Bush' and 'rapid increase of buildings on the Seven Sisters line of Road'. The result, as McAdam reported in 1839 and several other years, was 'a greatly increased thoroughfare [traffic] of omnibuses and other short stages'.[48] That was not pure gain, because although omnibuses brought in more revenue, they also wore out the roads.[49] Moreover the Commissioners considered it their duty to get rid of gates and bars when they could, particularly where the local traffic was heavy, such as Seven Sisters Road (1844) and Harrow Town (1847), so that they deliberately denied themselves opportunities of making good in revenue from short-stage traffic the sums they lost through the ruin of the long-distance carriers. The effect, moreover, was cumulative, because, as the report for 1848 says, the abolition of gates and side-bars 'has promoted the establishment of a large additional number of public conveyances . . . which have become the more necessary by the lengthened distance of general residences out of town'.

The rise in road traffic on newly-freed roads must have been the more provoking to the Commissioners because they never got any gratitude from the public for the gates they did remove : simply grumbles that they did not remove gates fast enough. Altogether the effects on the Metropolis Commission of the railways, of the growth of London, of increasing short-stage traffic, and of the interaction between them were complex and mostly unwanted. McAdam surveyed the scene in 1841. There had been much extra traffic and wear 'upon the upper portions of the Metropolis Roads during the past years, arising from the great increase in the number of public carriages . . . for the conveyance of the public to and from the several Railway

A 11662

WILLIAM JONES

Begs to acquaint his Friends that his OMNIBUS Starts at the following Hours, through BATTERSEA, CHELSEA, PICCADILLY, STRAND, and CHEAPSIDE.

FROM WANDSWORTH	From the Hercules Leadenhall Street
9 o'Clock in the Morning	Quarter-past 11, Morning
1 o'Clock, Noon	20 Min. past 3 Afternoon
Quarter-past 5, Evening	Quarter to 8, Evening

On Sundays.

FROM WANDSWORTH	FROM LEADENHALL STREET
Quarter-past 9, Morning	Half-past 11, Morning
Quarter-before 8, Evening	10 o'Clock, Night

Calls at the White Horse Cellar, at

6 Min. past 12, Morning	6 Min. past 4, Afternoon
	25 Min. past 8, Evening

Calling also at the King's Head, Ludgate Hill; Red Lion Strand; Black Horse Coventry Street; and Mr. Thomas's, Grocer, King's Road.

FARES

	s.	d.
From Wandsworth to the White Horse Cellar..	1s.	0d.
———— City..............	1	6
From the Raven, Battersea, to the White Horse Cellar..... .... }	0	9
———— City	1	3

Omnibus timetable, undated. The fares are high, by later standards.

Museum of London

Sir Goldsworthy Gurney's steam carriage ran from London to Bath and back in 1829 at 15 mph (average), and for three months in 1831 ran regularly between Gloucester and Cheltenham: the original caption is reproduced below.

Museum of London

The Guide or Engineer is seated in front, having a lever rod from the two guide wheels to turn & direct the Carriage & another at his right hand connecting with the main Steam Pipe by which he regulates the motion of the Vehicle. The hind part of the Coach contains the machinery for producing

the Steam, on a novel & secure principle, which is conveyed by Pipes to the Cylinders beneath & by its action on the hind wheels sets the Carriage in motion. The Tank which contains about 60 Gallons of water, is placed under the body of the Coach & is its full length & breadth – the Chimneys are fixed on the top of the hind boot & as Coke is used for fuel, there will be no smoke while any hot or rarified air produced will be dispelled by the action of the Vehicle. At different stations on a journey the Coach receives fresh supplies of fuel & water. The full length of the Carriage is from 15 to 20 feet, & its weight about 2 Tons. The rate of travelling is intended to be from 8 to 10 miles per hour. The present Steam Carriage carries 6 inside & 12 outside Passengers – the front Boot contains the Luggage.

An omnibus passing the Three Compasses Inn, Clapton, 1850.
Museum of London

The 'Enterprise' Steam Omnibus, 1833, built by William Hancock.
Museum of London

Stations, and by the extension of buildings in almost every direction along the lines of the several Roads. A large addition has therefore taken place in the number of horses kept in London, by which the carriage of hay, straw, corn, and green provender, paying only half the toll, has been much increased, and the roads have now to sustain, principally during the winter season, a severe additional and heavy return-carriage of manure, which pays no toll, and being principally in narrow-wheeled waggons, weighing from six to eight tons, causes the consumption of much [road-mending] material, and the expenditure of much labour.'

As the declining years went on there was no longer money for the ambitious standards of road-building and upkeep set by McAdam in his early mood of hopeful energy. With the loss of 'long thoroughfare', he reflected in his report for 1839, it might be prudent to use less expensive material than hard stone and to narrow the carriageway. Two years later there were no bidders for tolls in six out of fifteen districts and the Commissioners said they were 'continuing the system referred to in their former Reports, of abstaining from every expensive improvement'.

McAdam, when and where he could, gave up using granite, as for instance between Hammersmith and Hounslow, except through Brentford, in 1844. Market gardeners wanted it back in 1846, and in 1847 McAdam himself thought it would be desirable to use more granite in the area of Hampstead and Highgate 'in accordance with the memorial of the omnibus proprietors and inhabitants'. The omnibus proprietors were especially insistent. 'The unprecedented increase of omnibus traffic,' said McAdam, '. . . arising from competition and the great reduction of fares, has produced an earnest desire, on behalf of the proprietors, . . . for a hard, solid, and improved surface of Road, best calculated for the draught of such carriages, having only two horses, and conveying frequently upwards of twenty passengers.'

The bus owners even offered to pay extra tolls for a better road, but extra tolls – indeed, the very notion of tolls – ran against the tide of the times. In the same report in which he mentioned the extraordinary offer from the bus owners, McAdam also recorded the abolition of Paradise Row gate, Hornsey Lane bar, and a gate near Shoreditch Church, giving free access to 'the Metropolis' from that direction. This was the trend, as it had been ever since the foundation of the Commission, and the Commissioners had steadily sought to pass road upkeep and other services, such as watching and lighting, to parishes or, in the case of the watch, to the Metropolitan Police after that force was established in 1829. But if travellers didn't like paying tolls, neither did householders like paying rates, so that nothing the Commissioners did was likely to make them popular.

To the exuberant middle class of the 1840s, sweeping forward on the high tide of English economic supremacy from Reform to the Corn Laws, to the destruction of aristocratic privilege, to Free Trade and unfettered private enterprise, such bodies as the Metropolis Roads Commission stank of the old order and ought to be destroyed. 'These Metropolis Road Commissioners', wrote Viator to *The Times* on 23rd January 1845, 'have been eighteen years in official existence; they have a large and expensive house in Whitehall Place, containing about four times the room necessary for the business, and therefore their officers may probably have turned some of them into sleeping apartments.' Petitioners from St Pancras and Islington, writing two or three years after McAdam's death, made the point even more plainly. They referred to '. . . a self-elected Commission . . . appointed before the Reform Bill, and . . . the difficulty which is always attendant upon obtaining information from a body, upon which the rate or toll-payers are not represented.' They suggested, quite sensibly, putting the roads under the control of the Metropolitan Board of Works, 'a representative and open Board'. The nub of their complaint, familiar then and by no means unknown today, they set out in bold type:

'. . . it is, as your Petitioners submit, opposed to common sense to tax vehicles and horsemen to repair roads which, as civilized men in a civilized country, they have a right to use without let or hindrance of any kind whatever.'[50]

The numerous country trusts to which Sir James was surveyor did not have to cope with the consequences of the rapid expansion of a vast city. They did, however, have to cope with railways, and their experience, on a smaller scale, paralleled that of the Metropolis Roads Commission. To some the threat came early; to others, late. For some, the financial results were quickly disastrous, for others, more slowly, depending chiefly on whether they had local traffic to rely on as the long-distance traffic fell away. The Kneesworth & Caxton (Royston) Trust on the Old North Road, carrying the Edinburgh mails, had insufficient local traffic and was asking the Treasury for money even before the railways came, since most of it turned off at Royston. It was moribund in the forties. The Wadesmill Trust, on the other hand, was little affected by railways until the late forties, and the Hauxton and Dunsbridge Trust, with traffic for Cambridge Market to rely on, was very active in the early fifties, in spite of fast-falling revenue.[51]

As a trust drooped, the question of Sir James's salary would come before the trustees, or the question whether they could afford Sir James at all. Some, perhaps, decided they could not, for about ten of his surveyorships came to an end, for reasons of which we have no record, between 1837 and 1850, and he was not in all cases succeeded by his son, James Jr. Most of these trusts were in outlying districts away from his central citadel in London and the home counties, and it may be that he was not giving them enough attention. There are occasional complaints of that, or hints of it, in minutes that survive. To five trusts whose minutes still exist he made offers in the early fifties to reduce his salary: offers which none refused.* There were no doubt other cases also, so that in the last years of his life Sir James's income must have been dropping, and would no doubt have gone on dropping if he had lived longer.

---

* Wadesmill 1851 – £100 to £60; Cheshunt 1851 – £157 10s to £105; Newmarket 1852 – £50 to £40; Watton 1852 – £85 to £80; Welwyn 1852 – £90 to £70.[52]

He was not, however, easily discouraged. He took his son into partnership in several of his appointments and himself accepted, even sought, new work. As late as 1839, when it might be thought that he was sufficiently eminent to be approached by any trust wishing to employ him, he applied for the surveyorship to Essex 2nd District, and got it.[53] He took on a post at Stevenage and Biggleswade in 1845[54]: at Enfield Chase (Galley Corner) in 1846.

From this last post he was dismissed, late in 1850, for disobedience. His resignation in March 1851 is also on record, perhaps as a saver of face. After some hesitation, he had refused the Trustees' order to repair a road through Hatfield Park which the 2nd Marquess of Salisbury – 'general self-appointed dictator of the Hatfield district' according to Lord David Cecil[55] – was determined to close. It is evident from the detailed surviving records of the episode, which was complicated and long-drawn-out, that McAdam had no desire to fall foul either of the trustees or of the Marquess, being rather in the position of an embarrassed spectator of their quarrel. Forced to choose, however, he chose the Marquess, an old patron whom he knew well. On 28th January 1851 his appointment with the trust was cancelled and another surveyor appointed. It is a sad little story, coming late in a distinguished career which itself was drawing towards a rather melancholy end.[56]

What is perhaps saddest of all about Sir James McAdam's later life is that steam power, instead of destroying his hopes, might have realized them. Steam carriages, which in the 1830s were developing from inventors' experiments into practical road vehicles, might have been the salvation of the turnpike roads. As early as 1831, when a steam carriage service was already running between Cheltenham and Gloucester, McAdam and Telford, in evidence to a Select Committee, agreed that greater wear was caused on roads by horses' hooves than by the wheels of carriages, and in 1839 McAdam called the development of steam carriages 'the most fortunate circumstance that could occur for turnpike roads . . . as we should have upon the roads cylindrical wheels of $4\frac{1}{2}$ inches in width, and no horses' feet; and it is a well-known fact that horses' feet do more injury than the wheels of carriages'.[57]

It was not to be. The opposition to steam carriages, mainly from those interested in horse transport and in railways, was varied and powerful, and by 1865, except for traction engines travelling at four miles an hour behind red flags, steam power was driven from the roads. The roads then, instead of being adapted to mechanical transport, perhaps under some such national Board of Control as James McAdam envisaged and worked for, fell into their long 19th-century afternoon nap. As serious highways for long-distance transport, nobody from about 1840 until the end of the century was interested any more in roads, and getting anything like a national system of finance, planning and administration took much longer. Public interest in roads died with the stage coach and did not revive until the coming of the bicycle and the motor car.

Public interest died, too, in those who had created many of the best of the stage-coach roads. Less than a fortnight after J L McAdam died in 1836, Sir James McAdam wrote to the Duke of Richmond seeking his support for a committee to raise money for a monument, claiming that there would be plenty of supporters including the coach proprietors who 'have sent to me to say that they are only waiting an opportunity to come forward and indeed I think almost every Class would give their Mite'. The Duke endorsed the letter 'Happy to be of service' and took the chair at a meeting in London on 24th January 1837 at which a resolution was passed 'That as a mark of respect to his [J L McAdam's] Name and Memory, it will be desirable to erect a Monument in some Public situation.' A committee was formed including the Duke, two Marquesses, four Earls, four Viscounts, three Barons and ten other notables and a subscription list – maximum £10 – was opened.[58]

It failed miserably. About £518 was promised, and presumably paid, by 113 subscribers to the first appeal, but evidently very little more, for in February 1849 Sir James wrote to the Duke to say that 'after every matter is settled as regards the subscription' – presumably they had tried to return money to every subscriber they could trace – £57 still remained, which it was proposed to hand over to 'Bethlem Hospital'.[59]

The entire Committee subscribed at the top rate. So too did the Marquess of Salisbury, Lord Farnborough, Lord Holland, Earl Manvers, the Earl of Rosebery, E K Fordham the Royston banker, and Lewis Levy, toll farmer. Four members of the Gladstone family came in, either with five pounds or two, two more toll farmers – E and B Ambler – sent a guinea each and Captain Scobell, a pound. Among the other names turnpike trustees and clerks are recognizable, and no doubt there were some of the expected coach proprietors. What is striking is that the list seems to be essentially a personal one, made up chiefly of people who had known J L McAdam or members of his family or had had dealings with him or them professionally. To the general public, barely more than ten years after the period of his greatest fame and in an age fond of monuments, McAdam's name made no appeal.

Sir James McAdam died on 30th June 1852 in St John's Wood, aged 66. His son succeeded him in many of his appointments and then himself died, aged 33, in March 1853. Of the McAdam dynasty, so recently so flourishing, only John Loudon Jr and his nephew William remained, in obscurity in the West of England. As for Sir James, a distinguished public servant, it is difficult to find an adequate obituary and the Dictionary of National Biography tucks him away insultingly in a corner of his father's unsatisfactory entry.

Yet he was both a pioneer of modern management and one of the founders of a modern profession, and on the roads of the Home Counties and in London his monuments remain. He should be remembered above all, perhaps, in the townscape of 20th-century London, for Seven Sisters Road. It was the longest stretch projected, designed and built by the Metropolis Roads Commissioners, and it linked Central London with James's other trusts in Essex and beyond. This road system and many others which it was his life's work to care for are still in service. His name is in the shadows, but his works remain.

REFERENCES

Parliamentary Select Committees are referred to by year only, unless confusion might arise, as in the case of the SC on Metropolis Turnpike Trusts 1825. For full titles see list of sources, Appendix Three.

GC – Goodwood Correspondence (Correspondence of the 5th Duke of Richmond in the West Sussex Record Office).

1. Abstracts of Statements of Income and Expenditure of Turnpike Trusts 1836–52 (and thereafter, but in less detail, to 1861).
2. *Examiner*, exact date not given, quoted in support of a petition from the householders of St Pancras and Islington 1856. Guildhall Library.
3. Evidence to SC on Metropolis Turnpike Trusts 1825, 69.
4. J L McAdam, *Remarks* 8–9; Evidence to SC 1819.
5. SC 1819, Report; SC 1820, Report; SC 1821, Report.
6. DNB. See also F M L Thompson, *English Landed Society in the 19th Century*, 1963, 47 72 264.
7. (3) 3 5.
8. Commissioners of the Metropolis Turnpike Roads North of the Thames, Minute Books. Greater London Record Office.
9. First Report of the Commissioners of Metropolis Turnpike Roads North of the Thames, 1827, 3–4.
10. Letter to *The Times* 23i45. Report by James McAdam, Appendix 7 to (9) 20. 2nd and 4th Reports of Commissioners.
11. (9) 4.
12. 1st, 2nd, 4th Reports. Commissioners' Minutes 9 16ii27 2iii27 6viii27 – Minute Book No 1 (MRC 2). Commissioners' Letter Book (MRC17), letters of 19ii 8viii27, 9ii29, 5i 9i 23i30.
13. This paragraph and the two following are based on Appendix 7 to (9).
14. App 7 to (9), also Minute of 24x26.
15. Cf petition to Parliament 1857 by nearly 3,000 householders in Hampstead, Highgate, Holloway, Kentish Town, Camden Town &c. Guildhall Library.
16. Devereux, *John Loudon McAdam* 64. T C Barker and Michael Robbins, *History of London Transport* I 1963 13.
17. 2nd Report of Commissioners, 1828, 13.
18. (17). 22nd Report 1848.
19. 4th Report of Commissioners 1830, 4–5.
20. Benjamin Winstone, *Minutes of the Epping and Ongar Highway Trust*, privately printed 1891, Chapter IX.
21. (19).
22. (17) 9.

23. SC 1828, evidence of George Byng, 23; of Sir Henry Parnell, 22.
24. Minutes of the St Albans (Pondyards and Barnet) Trust, Herts RO.
25. (24) 6 13 16vi23; 3vi28. TP5 3–4, Herts RO.
26. For Telford's criticisms see various Reports of the SC on the Holyhead Road, also (23), evidence of Sir Henry Parnell 18–20, also (24) 25iv25 19v 3vi28.
27. (24) 29vi29 9x33.
28. (24) 29vi 26x29.
29. (24) 27vii33.
30. (24) 7xi36.
31. (24) 24xi52.
32. (23) Evidence of Sir H Parnell, 21.
33. (23) Evidence of James McAdam, 36.
34. DNB.
35. Letters of 24xii32 GC 1461/296; 14iii35, GC 1571/95; 21ii49 GC 1725/1206–7.
36. Letters of 27x34 GC 1477/364; 4x34 GC 1485/238; 12ii36 GC 1874/722.
37. Letter of 9i33 GC 1463/363.
38. Letters between McAdam and the Duke: 9 4 16ii 18iii33 GC 1463/429, 1486/337 1463/452 1464/546.
39. McAdam to the Duke 21 viii33 GC 1468/482 483.
40. McAdam to the Duke 1ii36 GC 1874/720.
41. McAdam to the Duke 6xii33 GC 1470/197.
42. SC 1839 33.
43. Devereux 135; Sir James McAdam to the Duke of Richmond 7xi35 GC 1579/459.
44. Sir James McAdam to the Duke of Richmond 28vii36 GC 1874/801.
45. GC 1487/24 40.
46. (16) 79.
47. McAdam to W C Kitchener 13xi46, Kneesworth & Caxton Minutes, Cambs RO.
48. On the rise of the omnibus, see Barker and Robbins (16) I, Chapter I.
49. Commissioners' Report 1837.
50. Petition from householders in St Pancras and Islington. Guildhall Library.
51. Minutes as (44) 8viii33, T/K/AM1, Cambs RO
52. Wadesmill Minutes, Herts RO, Hauxton and Dunsbridge (Cambridge Roads) Minutes, University Library, Cambridge, UL6014.
53. From Minutes in appropriate county ROs – see list of sources.

54. Letter of 5iii39, Essex RO D/TX9.
55. Lord David Cecil, *The Cecils*, 1973, 196.
56. See minutes of Enfield Chase (Galley Corner) Trust in Herts RO; letters from Sir James McAdam to Lord Salisbury at Hatfield House.
57. SC on Steam Carriages 1831, quoted in *Mr Gurney's Observations on Steam Carriages on Turnpike Roads*, 1832, 23. SC 1839 41.
58. McAdam to the Duke 8xii36 GC 1874/869; leaflet 'Proposed Monument to . . . John Loudon McAdam' GC 1725/1207.
59. McAdam to the Duke 21ii49 GC 1725/1206–7.

VIII

From McAdam to the Motor Car: the Roads of Victorian England

by W J Reader and Lauraine Dennett

THE MCADAMS WERE Scots and they were innovators: on neither count were they likely to be popular in the rural England of their day. Moreover they thrust their innovating Scottish noses into areas particularly resistant to change. By specifying a particular technique of road construction they challenged the conventional wisdom of centuries: scepticism was not easily, nor in all cases permanently, overcome. By insisting on professionalism in surveyors they interfered with local patronage valued both by bestowers and receivers, and they threatened the easy profits of contractors. Above all, their demand for the consolidation of turnpike trusts and for central control of the roads affronted upholders, high and low in the social scale, of the ancient tradition of local administration of local affairs.

McAdam to the Motor Car

On the other hand there was a strong demand for what the McAdams had to offer : better roads for faster travel. Moreover the current of the times was with them. Change was all about them : in technology, symbolized by the rise of steam power; in politics, by Reform, Chartism, Free Trade; in literature and music, by romanticism; in manners and morals, by the Evangelical Movement on the one hand and by the Oxford Movement on the other. In this atmosphere it was to be expected that the pressure of new demand would bear down the resistance opposed to it and that the McAdams, like other great innovators of the day in all the fields mentioned, would eventually see their system become a part of the nation's stock of received ideas.

The McAdams were unlucky. They were caught in one of the characteristic situations of an expanding economy; the clash of competing technologies, in this case the technology of road transport and the technology of transport by rail. To the first they made the vital contribution of McAdam's system of road construction and management. To the second they had nothing to offer. If steam carriages had been allowed to develop, the case would have been altered and the successful competition of road against rail might have been brought forward a hundred years or so, but when the last road-making McAdam – William Jr – died, the age of mechanical road transport had been postponed : it was still forty years in the future.

The consequence was that the road administrators, if that is not too grand a description, of the country were able to fall back, with an almost audible sigh of relief, into the old ways from which the Scotch invasion had so rudely disturbed them. Indeed they were able to fall right through the turnpike trusts, those odious 18th-century innovations, and back towards still older ways, for in the brisk new liberal age of free trade, steam and reform nobody had a good word to say for paying tolls to non-elected, self-selected bodies for the right to use the roads, and after everybody took to travelling and sending their goods by train there was no longer any pressing necessity to find money for road upkeep, improvement and management. The

199

turnpike trusts disappeared with increasing rapidity in the sixties and seventies, and the parishes, Dogberry, Verges and all, came into their own again.

It is almost impossible, nowadays, to visualize the country roads of Victorian England, except in faded photographs with gossiping couples and groups standing unconcerned in the middle of the road and children playing near them. Long-distance traffic had all been captured by the trains and local traffic, except in market towns on market day, was not likely to be heavy. Flora Thompson, looking back from the 1930s to the 1870s or 1880s, describes a road near Buckingham, probably A421, like this:

> 'Although it was a main road, there was scarcely any traffic, for the market town [Buckingham] lay in the opposite direction along it, the next village was five miles on, and with Oxford there was no road communcation from that distant point in those days of horse-drawn vehicles. To-day, past that same spot, a first-class, tar-sprayed road, thronged with motor traffic, runs between low, closely trimmed hedges. . . . At that time it was deserted for hours together. Three miles away trains roared over a viaduct, carrying those who would . . . have used the turnpike. People were saying that far too much money was being spent on keeping such roads in repair, for their day was over; they were only needed now for people going from village to village.'[1]

These were scarcely circumstances propitious to the success of the McAdams' chief strategic aim in highway policy: to bring roads under larger local units, themselves under some form of central control. It had never been close to realization, and in the last years of Sir James McAdam it disappeared entirely from practical politics, though as late as 1849 he lamented to his old friend the Duke of Richmond: 'Our poor Road Bill is gone'[2], suggesting that even then he still had hope.

Over the years between the 1830s and the 1890s an incoherent jumble of authorities, most of them tiny, arose to administer the roads which had chiefly been the responsibilities of the turnpike trusts : that is, the roads across country outside the large towns. The most ancient, the 'highway parish', survived until the end of the century, ever more entangled within a forest of 'highway districts' (1862), 'rural sanitary authorities' (roads somehow got mixed up with public health) (1872), 'county authorities' – the justices in quarter sessions (1878), County and County Borough Councils, all of which existed alongside, over or around the parish, in some places superseding it, in others, not. The Local Government Act of 1894 finally, though not speedily, got rid of highway parishes, replacing them with rural (and urban) district councils, though even then there remained six different kinds of highway authority : County Councils, County Borough Councils, Urban District Councils, Rural District Councils, Borough Councils and London authorities. Over the whole scene the President of the Local Government Board held sway, having in 1872 taken over from the Home Office such responsibility as the central government possessed for the administration of the highways. The Board consisted of the President, a Parliamentary Secretary and a number of high officials. Up to 1912 it had never met, but the presidency provided a useful appointment for middle-rank politicians like John Burns (1858–1943), President from 1905 to 1914.[3]

In the general mid-century decline of the roads the professional road surveyor, as trained under McAdam's 'system', seems almost to have died out, as the McAdams themselves died out. Even in the hey-day of the turnpike trusts, the professional surveyor was always in danger of being discarded as an expensive luxury – as Loudon McAdam was discarded by Sheffield & Glossop in 1822, William by Truro in 1829, and Christopher by Exeter in 1837[4] – and as the railways put an end to the trusts no such appointment was likely to be contemplated by parishes or highway districts. Parishes were authorized by the Act of 1835 to combine resources to hire a surveyor on a group basis, but they rarely did so, preferring instead to remain uncombined and to elect unqualified amateurs as 'overseers'. As late as 1903 a Departmental Committee on

Highways heard that small local authorities, successors to the highway parishes, would 'appoint gardeners, artisans, clerks of works and all sorts of people who know nothing about roads' as surveyors at salaries ranging from £40 to £150 a year : too little to attract a well-qualified man.[5]

As time went on, nevertheless, influences increasingly came into play to restore surveyors of roads to something like the standing claimed for them by McAdam, although for surveyors generally the main line of professional advance was in the direction of service to property owners, including railway companies, in which the construction and maintenance of roads was only incidental.[6] In towns, local Boards of Health, set up from the fifties onward and turned into urban district councils in 1894, took over the appointment of road surveyors from the parishes. The officials they appointed, chiefly civil engineers and architects, had much else to look after besides roads, but their professional skill – supported, no doubt, by adequate finance – rapidly brought town roads into much better condition than roads in the surrounding countryside.

Finance lay at the heart of the matter for, as J L McAdam had often pointed out and the turnpike trusts had discovered, a surveyor with a properly professional approach to his work and the appropriate social standing was expensive. In 1878 the Highways and Locomotives (Amendment) Act provided that roads 'disturnpiked' after 31st December 1870, and other roads on appeal, should be designated 'main roads', and then half the expense of their upkeep should be found by the 'county authority' from the county rate, leaving only half to be raised by the parish or highway district. That lifted from local shoulders some of the burden, long complained of, of roads of more than purely local importance, but there was a price in local independence to be paid. Before the county would meet its share of the bills, they would have to be certified by a surveyor appointed – as J L McAdam, over fifty years earlier, had suggested – by the county authority.

Thus the County Surveyor, trained perhaps as a civil engineer or architect or with a firm of chartered surveyors –

their professional institution was incorporated in 1868 and chartered in 1881[7] – began to appear upon the country roads. After 1888 the new County Councils and County Borough Councils took over the whole cost of maintaining main roads, even those through towns unless the town councils elected to remain in control, and they could contribute, at their discretion, to the cost of secondary roads as well. With that enlargement of the County's functions, the County Surveyor blossomed. He initiated and planned work on the many miles of main road found in disrepair when they were taken over. He contracted for supplies, making his favour much sought after.[8] He evaluated the repairs for which the County Council contracted with the subordinate highway authorities. After 1894 he supervised a team of district surveyors, making a leisurely round every few weeks, perhaps in his own carriage. His social position was unquestionable : the holder of a county post, like his brethren the City and Borough Engineers and Surveyors, regarded himself and was accepted as a professional man. To accommodate him, the Association of Municipal and Sanitary Engineers and Surveyors, founded in 1873, changed its name in 1891 to the Association of Municipal and County Engineers and Surveyors.

The early county surveyors found themselves faced with the early cyclists. A road good enough for a farmer's cart, or even for his horse and trap, was by no means good enough for a bicycle, which needed a road built to the highest standards, just as the fast stage coaches did, and for the first time since the stage coaches disappeared there came, from the cyclists, a powerful demand for better roads. Once again, as in J L McAdam's early days, there was public interest in the roads and some willingness to see money spent on them. An obstacle to progress was the multiplicity of highway authorities. Evidence given to the Departmental Committee of 1903 showed seventy-two authorities along the road between London and Carlisle and a dozen on eighteen miles of road running westward from London. The result, even on major roads, was an alarming variation in the quality of the surface, even on major

roads, as the traveller passed from the care of one authority
into that of another. By-roads, even after 1888, might still be
repaired with lumps of rock of a size that would have appalled
J L McAdam, dumped at intervals along the road, and left for
the wheels of traffic to flatten. Occasionally a rural authority
might borrow a roller from the County Council, but that was
by no means standard practice and few such authorities could
afford road-making equipment of their own.

Some time after the turn of the century an enterprising Scot,
Harry Inglis, having published *The 'Contour' Road Book of
Scotland*, published a similar work for England and Wales. He
was aiming chiefly at cyclists – even as late as 1914 – and
therefore paid great attention to hills (hence the title) and to
road surfaces. The main road between London and Bath, one
of the principal roads of the country, he calls 'a splendid road
throughout', and yet : 'Near Reading the road is often not in
very good condition, and after rain it is very heavy' – words
which would have had an all too familiar ring to J L McAdam
or James in the 1820s. Of one of J L McAdam's favourite
roads, the one across the Somerset levels towards Bridgwater,
he says : 'After Cross the road is almost level, but with bumpy
surface, to Highbridge, then better approaching Bridgwater.'
He is complimentary about the road through Epping to
Bishop's Stortford, in the midst of James's country, which has
'a fine surface', and from Bishop's Stortford to Cambridge 'the
surface is fine the whole way, and the gradients are hardly
perceptible after Littlebury'. Cyclists, being more vocal than
coach-horses, were more outspoken about gradients.[9]

Cyclists were especially vocal through the Roads Improvement Association, formed jointly by the two national cycling
clubs in 1886. For ten years the Association concentrated on
publishing helpful handbooks such as *Hints to Country Roadmen*
and *Our Roads and How to Treat Them*, which were circulated
in thousands to road surveyors and the general public. When
motor cars were allowed to run freely on the roads, after 1896,
the Automobile Club of Great Britain and Ireland, later the
RAC, co-operated with the RIA in mounting a campaign for

improving the roads which developed into a demand for central financing and hence for some form of central control. In 1909 Lloyd George put a petrol tax into his Budget and increased the fee for motor vehicle licences, the intention behind both measures being to provide money to improve the roads. Funds centrally raised had of course to be centrally administered, and at last, in 1910, something rather like J L McAdam's proposal of 1811 (see p. 35), for 'measures, under the authority of Parliament, to oblige all trustees, commissioners and way-wardens to conform to such regulations in making and repairing roads as shall be laid down', came into effect. A Road Board was set up: the first national road authority in Great Britain since Roman times.[11]

The RIA and its allies had meanwhile been seeking means to alleviate the dust nuisance, that pest of the old white roads. Always tiresome, it became intolerable when the motor car arrived, as Kenneth Grahame reminded readers of *The Wind in the Willows*:

'The "poop-poop" rang with a brazen shout in their ears, they had a moment's glimpse of an interior of glittering plate-glass and rich morocco, and the magnificent motor-car, immense, breath-snatching, passionate, with its pilot tense and hugging his wheel, possessed all earth and air for the fraction of a second, flung an enveloping cloud of dust that blinded and enwrapped them utterly, and then dwindled to a speck in the far distance, changed back into a droning bee once more . . .'[12]

One way of getting rid of dust, technically possible by 1900, was to make roads of concrete. Another was to break stone small, in the McAdam manner, and bind it with tar to form 'tarmacadam'. That material had been used in towns, but not in the countryside, as early as 1832 and quite widely from the seventies on. In the early years of the new century experiments were going on, with the encouragement of the RIA, to find some relatively cheap means of eliminating dust from rural roads and providing a smooth and water-proof surface fit for bicycles and cars.

Among those interested, about the turn of the century, in the use of tar on roads was the County Surveyor of Nottinghamshire, E P Hooley, who was observant, inventive and enterprising. On a road near Denby Iron Works in Derbyshire in 1901 he noticed that a barrel of tar had burst and run over a road lightly covered with slag from the furnaces, producing a dust-free, hard-wearing surface. What had been done accidentally could be repeated purposely, and Hooley proceeded to do it. The eventual result, considerably elaborated from the original inspiration, was a new road material and a new company to produce it – Tar Macadam (Purnell Hooley's Patent) Syndicate Limited – which in a moment of word-coining brilliance was renamed Tarmac Limited.

If concrete had been widely used on roads instead of tar; if Hooley's original name for his company had been replaced by some word other than Tarmac – Tarslag? Hooleytar? Tarnell? – it is fair to assume that McAdam's name would have dropped even more deeply into obscurity. By pure chance, instead, it has been preserved in folk memory by an abbreviation built into the trade name of a product which has not the faintest connection with the McAdam family.

It is right that the family's name should be widely remembered, however quirkishly. Their ideas on road management, driven underground by the temporary triumph of the railways, came to the surface again with the revival of the roads. That ideas such as the appointment of County Surveyors and the establishment of a national Road Board had originated with J L McAdam had by then been forgotten. Mr Purnell Hooley, quite unwittingly, ensured that justice was done in the end.

REFERENCES

1. Flora Thompson, *Lark Rise to Candleford*, World's Classics Edn 22.
2. Sir James McAdam to the Duke of Richmond 24iv49 GC 1588/1225.
3. Highway Act 1862; Public Health Acts 1872 and 1875; Highways and Locomotives (Amendments) Act 1878; Local Government Acts 1888 and 1894; W Rees Jeffreys, *The King's Highway* 7.
4. Sheffield-Glossop Trust Minutes 13vi22, Sheffield City Libraries; Truro Trust Minutes 14i33; Exeter Trust Minutes 23viii37.
5. Minutes of Evidence, Departmental Committee on Highways, BPP 24 1904, evidence of William Rees Jeffreys.
6. F M L Thompson, *Chartered Surveyors, the Growth of a Profession* 139.
7. Geoffrey Millerson, *The Qualifying Associations* 239.
8. J B F Earle, *A Century of Road Materials* 7–9.
9. Harry R G Inglis, *The 'Contour' Road Book of England*, Western Division, 1912–13, Route 845–6; South Eastern Division, 1913–14, Routes 517 537–8.
10. Rees Jeffreys as (3) Ch 1 and 2.
11. As (10) 23–45.
12. Kenneth Grahame, *The Wind in the Willows*.

Appendix Two: Family Connections

Names of road surveyors italicized.

1. The Direct Line:

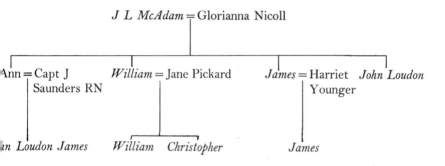

J L McAdam = Glorianna Nicoll

Ann = Capt J Saunders RN   *William* = Jane Pickard   *James* = Harriet Younger   *John Loudon*

in Loudon James   *William*   *Christopher*   *James*

2. Collateral Lines:

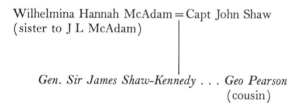

Wilhelmina Hannah McAdam = Capt John Shaw
(sister to J L McAdam)

Gen. Sir James Shaw-Kennedy . . . *Geo Pearson*
(cousin)

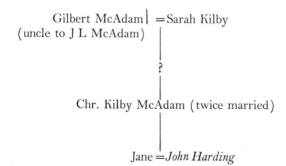

Gilbert McAdam = Sarah Kilby
(uncle to J L McAdam)

?

Chr. Kilby McAdam (twice married)

Jane = *John Harding*

215

Appendix Three: Source Material

A. Unpublished Documents:

Correspondence:

Correspondence of Sir James McAdam with the Duke of Richmond, in the West Sussex Record Office, Chichester.

Correspondence of Sir James McAdam with the Marquess of Salisbury, at Hatfield House.

Minutes and other Papers of Turnpike Trusts, in the appropriate County Record Offices unless otherwise indicated:

Berkshire: Fyfield

Cambridgeshire: Arrington, in Cambridge University Library
Cambridge & Ely – North District
North-West District
South District
Hauxton & Dunsbridge (Cambridge Roads), in Cambridge University Library
Kneesworth & Caxton (Royston, Old North Road)
Newmarket Heath

Cornwall: Launceston
Truro

Devon: Exeter

Essex: Epping & Ongar
Essex & Hertfordshire (Hockerill)

Hertfordshire: Cheshunt
Galley Corner (Enfield Chase)
Lemsford Mill & Welwyn, at Hatfield House
St Albans (Pondyards & Barnet)
Sparrows Herne
Stevenage & Biggleswade
Wadesmill
Watton

Middlesex: Metropolis Roads Commission

Somerset: Bath
High Ham & Ashcott
Shepton Mallet
Wells

Wiltshire: Black Dog (plan of new Warminster-Bath road: see p. 153 above).

Appendix Three

Frome (plan of the roads only)
Melksham (Telford's proposal for improving the
London road near Bath, p. 150 above).
Yorkshire: Sheffield & Glossop, in Sheffield Central Library.
PhD Theses: B J Buchanan: Capital Formation in North Somerset
1750–1830, University of Bristol, 1979.
Robert H Spiro Jr: John Loudon McAdam, Colossus
of Roads, University of Edinburgh, 1950.

B. Publications:

Albert, William: *The Turnpike Road System in England*, Cambridge
University Press, 1971.
Atthill, Robin: *Old Mendip*, David & Charles, 1964.
Ballen, Dorothy: *Bibliography of Road-Making and Roads in the
United Kingdom*, P S King, 1914.
Barker, T C and Robbins, Michael: *A History of London Transport* I
The Nineteenth Century, George Allen & Unwin, 1963.
Betts, John: *Betts's British Stage Coach, Van, and Wagon Directory
for 1831*, London, 1831.
Buchanan, Angus and Cossons, Neil: *Industrial Archaeology of the
Bristol Region*, David & Charles, 1969.
Cary, John (Compiler): Cary's New Itinerary, 11th Edn, London,
1828.
Chandler, Alfred D, Jr: *The Visible Hand*, Harvard University
Press, 1977.
Chapman, Herbert T: *Reminiscences of a Highway Surveyor*,
Maidstone, 1932.
Copeland, John: *Roads and their Traffic 1750–1850*, David &
Charles, 1968.
Cossons, Arthur: *The Turnpike Roads of Nottinghamshire*, Historical
Association Leaflet No 97, 1934.
'Devereux, Roy' (Mrs D Pember): *John Loudon McAdam*, Oxford
University Press, 1936.
Dyos, H J and Aldcroft, D H: *British Transport, An Economic
Survey from the Seventeenth Century to the Twentieth*, Leicester
University Press, 1969.
Earle, J B F: *A Century of Road Materials*, Basil Blackwell, 1971.
Blacktop: a History of the British Flexible Roads Industry, Basil
Blackwell, 1974.
Emmison, F G: *Turnpike Roads and Toll Gates of Bedfordshie*,
Bedfordshire Historical Records Society, Aspley Guise, 1936.
Flower, E F: *The Stones of London, or Macadam v. Vestries*,
London, 1880.

Macadam

Gentleman's Magazine, January 1837 (Obituary of J L McAdam); August 1852 (Sir James' McAdam's death); April 1853 (James McAdam Jr's death); 1861 pt ii, 455 (Obituary of William Jr).

Gibb, Sir Alexander: *The Story of Telford*, Alexander Maclehose, 1935.

Glen, Alexander: *The Highway Acts 1862–1878*, 5th Edn, 1879. *The Powers and Duties of Surveyors of Highways*, London, 1880.

Good, Ronald: *The Old Roads of Dorset*, Bournemouth, 1966.

Harper, C G: *The Bath Road, The Exeter Road*, both Chapman & Hall, 1899.

Hewitt, H: 'The Maintenance of Macadamised Roads', Surveyors' Institution, *Transactions* XXI (1888–9), 19–65.

Holgate, Jerome B: *American Genealogy*, Albany, NY, 1848.

Hooley, E P: 'Maintenance of Main Roads and County Management', Association of Municipal and Sanitary Engineers and Surveyors, *Proceedings* XVI (1890), 174–201.

Hughes, S: *A General Survey of the principal Metropolitan Roads*, prefixed to Henry Law's *Rudiments* (qv below).

Inglis, Harry R G: *The 'Contour' Road Book of England* – Western Division (including Wales) 1912–13; South-East Division 1913–14, both Gall & Inglis, London and Edinburgh.

Jackman, W T: *The Development of Transportation in Modern England*, 2 vols, Cambridge University Press, 1916; reprinted, Frank Cass, 1962.

Jeffreys, William Rees: *The King's Highway*, Batchworth Press, 1949.

Law, Henry: *Rudiments of the Art of constructing and repairing Common Roads*, London, 1850, and see under Hughes, S above.

Lewis, Michael: *The Navy in Transition*, Hodder & Stoughton, 1965.

Luton Museum: *The Turnpike Age*, Luton, 1970, and later reprints.

McAdam, J L: *Remarks on the present System of Road Making*, 7th Edn, revised, with the Report from the Select Committee of 1823 and extracts from the evidence, Longmans, London, 1823. *Observations on the Management of Trusts*, Longmans, 1825. *Narrative of Affairs of the Bristol District of Roads from 1816 to 1824*, Bristol, nd (1825).

MacDermot, E T: *History of the Great Western Railway*, 2 vols, London, 1927.

McKie, H U: 'Tar Macadam and Concrete Macadam Pavement for Roadways, and Tar Pavement and Cement Pavement for Footways'. Association of Municipal and Sanitary Engineers and Surveyors, *Proceedings* X (1883–4), 53–65.

Macmahon, K A: *Roads and Turnpike Trusts in Eastern Yorkshire*, East Yorkshire Local History Society, 1964.

Mahaffy, R P: *Highway and Road Traffic Law*, London, 1935.

Mathias, Peter: *The First Industrial Nation*, Methuen, 1969.

Maud, F H: *The Hockerill Highway*, Colchester, 1957.

Middlemas, R K: *The Master Builders*, Hutchinson, 1963.

Mogg, Edward (ed): *Paterson's Roads*, Longmans and others, 18th edn, 1826.

Nash, F W: *A Practical Guide to Surveyors of Highways*, London, 1836.

New-York Historical Society: Orderly Book of the Three Battalions of Loyalists commanded by Brigadier-General Oliver De Lancey 1776–1778. New-York Historical Society, 1917.

Oke, George C: *The Laws of Turnpike Roads*, 2nd Edn, Butterworth, 1861.

Ordish, T F: *History of Metropolitan Roads*, Appendix H to the Report of the London Traffic Branch of the Board of Trade 1910, Cd 5472/1911 – BPP 1911 24.

PARLIAMENTARY PAPERS:

Reports from Select Committees on Highways and Turnpike Roads:
    1811, 1810–11 3 855.
    1819, 1819 5 339.
    1820, 1820 2 301.
    1821, 1821 4 343.

Reports from Select Committees on the Roads from Holyhead to London:
    1817, fifth Report, 1817 3.
    1819, sixth Report, 1819 5.
    1820, second Report, 1820 3.
    1822, fourth Report (Road through England from North Wales), 1822 6.

Reports from other Select Committees:
    On Mr McAdam's Petition, relative to Road Making, 1823 – 1823 5 53.
    On Labourers' Wages, 1824 – 1824 6 401.
    On Metropolis Turnpike Trusts, 1825 – 1825 5 167.
    On the Whetstone and St Albans Turnpike Trusts, 1828 – 1828 4 255.
    On Turnpike Trusts, 1839 – 1839 9 369.

Reports from Commissioners of Metropolis Turnpike Roads:

1st 1826–7	7	23	12th 1837–8	35	125	23rd 1849	27	265
2nd 1828	9	23	13th 1839	29	577	24th 1850	30	247
3rd 1829	5	21	14th 1840	27	455	25th 1851	29	279
4th 1830	15	135	15th 1841	12	249			

5th	1831	12	1	16th	1842	25	219
6th	1831–2	23	541	17th	1843	29	305
7th	1834	40	211	18th	1844	31	287
8th	1834	40	223	19th	1845	27	181
9th	1835	36	319	20th	1846	24	375
10th	1836	36	383	21st	1847	33	347
11th	1837	33	213	22nd	1847–8	37	1

Other Reports, Returns &c:

Mr McAdam – Return of Salaries, Gratuities, or Remunerations received by Messieurs McAdam 1818–1825 – 1825 (248) – Turnpike Trusts 1814–1833, Reports & Returns Vol 10 pt 2.

Abstract of Statements of Income and Expenditure of Turnpike Trusts:

1836	47 297	1844 42 343	1852–3 97 335
1837	51 291	1845 41 447	1854 66 1
1837–8	46 97	1846 40 409	
1839	44 299	1847 44 421	
1840	45 391	1847–8 51 359	
1841	27 83	1850 49 379	
1842	37 247	1851 48 325	
1843	48 313	1852 44 305	

Letter from Sir James McAdam to Mr Cornewall Lewis, relative to Receipts or Expenditure for Highways in England. 1847–8 51 345.

Report and Minutes of Evidence of the Departmental Committee on Highways appointed by the President of the Local Government Board, Cd 1793 1794 of 1904, BPP 1904 24.

Report of the London Traffic Branch of the Board of Trade 1910 – Reports from Commissioners &c 1911 34.

Parnell, Sir Henry: *A Treatise on Roads*, 2nd Edn, enlarged, Longmans 1838.

Paterson, D: *Paterson's Roads*, see under Mogg, Edward (ed).

Pawson, Eric: *The Turnpike Trusts of the Eighteenth Century*, Research Paper 14, School of Geography, University of Oxford, November 1975.

Pember, Mrs D: see under 'Devereux, Roy'.

Phillips, John E: *The Local Government Act, 1888, as it Affects the Highways*, London 1888.

Pollins, Harold: *Britain's Railways*, David & Charles 1971.

Pratt, Edwin A: *A History of Inland Transport and Communication*, David & Charles Reprints 1970.

Rees-Jeffreys, William: *The King's Highway*, Batchworth Press, 1949.

Robertson, Alan W: *Great Britain – Post Roads Post Towns and Postal Rates 1635–1839*, privately printed, 1961.

Robertson, William: *Old Ayrshire Days*, John Menzies, Glasgow & Edinburgh, 1905.

Robinson, James: 'County Management of Main Roads'. Association of Municipal and County Engineers and Surveyors, *Proceedings* XVIII (1892), 169–198.

Rowe, M M and Jackson, A M: *Exeter Freemen 1266–1967*, Devon & Cornwall Record Society, Exeter, 1973.

Savage, Christopher: *An Economic History of Transport*, Hutchinson, 1966.

Sergeant, W R and Penrose, D G (eds): *Suffolk Turnpikes*, Suffolk Documents I, Ipswich and East Suffolk Record Office, Ipswich, 1973.

Searle, Mark (ed): *Turnpikes and Toll-Bars*, 2 vols, Hutchinson, n d.

Sheldon, Gilbert: *From Trackway to Turnpike: an Illustration from East Devon*, Oxford University Press, 1928.

Smith, Peter: *The Turnpike Age* – see under Luton Museum.

Society for the Diffusion of Useful Knowledge: *The British Almanac*, annually from 1828, Charles Knight, London.

Spiro, Robert H Jr: 'John Loudon McAdam in Somerset and Dorset', *Notes and Queries for Somerset and Dorset*, August 1956. 'John Loudon McAdam and the Metropolis Turnpike Trust', *J Transport History* II, 4 Nov 1956.

Spooner, Walter H (ed): *Historic Families of America*, New York, 1907.

Staffordshire County Council: *Staffordshire Roads 1700–1840*, Local History Source Book No 1, 1968.

Temple, John: *Darlington and the Turnpike Roads*, Darlington Public Library Local History Publications No 1, 1971.

Thompson, F M L: *Chartered Surveyors, the Growth of a Profession*, Routledge & Kegan Paul, 1968.

Thropp, James: *Repairs of Roads. Suggestions to Surveyors of Highways for the Management of Main Roads and Other Highways.* London and Lincoln 1887.

Toll Reform Association: Petition to the House of Commons . . . also Opinions of the Press on the Toll Bar Nuisance 1858.

Tonkin, W G S: *The Lea Bridge Turnpike and the Wragg Stage Coaches*, Walthamstow Antiquarian Society Monographs (New Series) No 14, 1974.

Tymms, Samuel: *Camden's Britannia epitomized and continued*, with Maps, 7 vols, London, n d.

Urwin, A C B: The Isleworth, Twickenham and Teddington Turnpike 1767–1872. The Hampton-Staines Turnpike 1773–

1859. Borough of Twickenham Local History Society, papers 3 and 11, October 1965 and March 1968.

Webb, Sydney and Beatrice: *The Story of the King's Highway*, Longmans Green, 1920.

Webster, Norman W: *The Great North Road*, Adams and Dart, Bath, 1974.

Winstone, Benjamin: *Minutes of the Epping & Ongar Highway Trust 1769–1870*, privately printed, 1891.

Wright, R S and Hobhouse, H: *An Outline of Local Government and Local Taxation in England and Wales*, London, 1884.

TABLE I : The Invaders' Progress – Surveyorships held by Members of the McAdam Family, 1816–1826

Figures following place names = Miles.

	J L McAdam	William	James	Loudon
1816	Bristol 179			Before 1823 : exact
1817			Epsom 21	years not known :
1818		Devizes 20	Cheshunt 17	
		Market	Royston 15	Ledbury 50
		Lavington 21	Wadesmill 29	Broomyard 50
		Melksham 10	Staines 9	Ludlow (part) 35
		Salisbury 30	Huntingdon	Hundred House[2] 17
		Newbury 20	Somersham 20	Kidderminster
		Reading[1] ?		(part) 30
1819		Black Dog[2] 20[6]	Oxford 20	Wellington 25
		Warminster 3	Egham West 8	Worksop-
		Southampton 9	Hertford	Retford[3] 9
		Popham Lane 12	Bridge 30	Worksop-
			Oakingham 20	Mansfield[3] 12
			Farnham 20	Pontefract 19
			Wycomb 14	Chester-
				Wrexham 11
				Chester-Northop 11
1820		Exeter 150	Welwyn 20	Flint
		Frome 42[6]	St Albans 11	Holywell ⎫
		Westbury 20	Beaconsfield 8	Mostyn ⎬ 40
		Andover 30	Boston 28	Scarborough-
		Winchester–	Chichester 19	York 46
		Alton 20	Whetstone 8	York-
		Winchester 22	Godalming 20	Northallerton 33
			Hatfield	Buxton-
			Reading 56	Macclesfield 13
			Stamford 42	Stretford 4
1821		Barnstaple 68	Sleaford 26	Between 1823 and
			Grantham 15	1825 :
			Sparrows	
			Herne 27	Cheltenham
			B/Edmunds	Litchfield
			–Sudbury 22	Shepton Mallet
			Henley and	
			Oxford 27	
			Watton 12	
1822		Totness 40	Kettering 12	
			Hockerill 27	
			Abingdon	
			–Fyfield 45	
			Higham	
			Ferrers 12	
			Spilsby 20	
			Stamford Hill 20	

J L McAdam	William	James	James Shaw
		Sevenoaks 11	
		Tonbridge 18	
		Wrotham	
		Heath 17	
			Edinburgh (1819–21)
1823	Plymouth 30	Lynn Districts 94	
		Pinner 10	
		Kingston	
		District 8	
		Newmarket	
		Heath 13	
1824	Cerne Abbas ⎫ Sturminster[4] ⎬ Kingsbridge ⎭	Cambridge & Ely (NW) 18	
1825 Carse of Gowrie 22[5]	Truro[6] Launceston[6]		273
1826 Bath 49[6]		Metropolis Roads 131	
	New Roads[7]		

Notes:

[1] Of the Reading Trust, William noted: 'Acted for my Father, and received from 1818 to 1823 – £200 to £250.' No trust is listed under this name in the returns to the Select Committee of 1821 and it is impossible to identify it.

[2] Black Dog and Hundred House are the names of inns where the trustees were accustomed to hold their meetings.

[3] These are all described by Loudon as new roads, just built or building, when he gave evidence to the Select Committee of 1823 on his father's petition for a Parliamentary grant, and the appointments may have been temporary.

[4] These were temporary appointments – Cerne Abbas, May 1824 to May 1826; Sturminster, July 1824 to July 1826; Kingsbridge three years from September 1824.

[5] Nine miles added under a new Act in 1832.

[6] During this period William Jr was his grandfather's colleague at Bath and his father's at Southampton, Black Dog, Frome, Cerne; Christopher was his father's (William Sr's) colleague at Exeter, Truro and Launceston; London was his father's colleague at Bristol.

[7] William listed 'New Roads' as follows: Haldon, 4½ miles; Taphouse, 6 miles; Minehead, 14 miles; Torbay, 3 miles; Modbury, 14 miles. Haldon and Taphouse were in the territory of the Exeter trust, Minehead in the territory of the Minehead Trust, Torbay probably in the territory of Totness Trust and Modbury, 12 miles East by South of Plymouth, probably in the area of the Plymouth Trust. All these roads were built under special Acts.

General: The names of the trusts are as given by the McAdams in evidence to the Select Committees of 1823 and 1825. Not all can be identified in the returns made to the Select Committee of 1821 or in the returns made under the Act of 1833.

See also Appendix One, List of all Trusts known to have been held by members of the McAdam Family.

Sources: William's Trusts: Return of Salaries and Gratuities received by the Messieurs McAdam, 1818–1825; James's and Loudon's Trusts, James's and Loudon's evidence to the Select Committee on Mr McAdam's Petition, 1823, modified by their evidence in the Return of Salaries and Gratuities.

224

TABLE II: James's Progress, 1817–1826

1817	December	Epsom
1818	September	*Cheshunt*
	November	*Kneesworth and Caxton**
	December	*Wadesmill*
		Staines
		Huntingdon/Somersham
1819	March	Oxford
	June	Egham West
	September	Hertford Bridge
	October	Oakingham
	November	Farnham
	December	Wycomb
1820	January	*Welwyn*
		St Albans
		Beaconsfield
	February	Boston
	April	Chichester
	June	Whetstone
	July	Godalming
	October	Hatfield and Reading
		Stamford, 5 roads
1821	February	Sleaford
		Grantham
	March	*Sparrows Herne*
	June	B/Edmunds Sudbury
	July	Henley and Oxford
	October	*Watton*
1822	January	Kettering
	February	*Hockerill* (Bishops Stortford)
	March	Abingdon/Fyfield
	August	Higham Ferrers, Northants
	October	Spilsby, Lincs
	December	Stamford Hill
		Tonbridge
		Sevenoaks
		Wrotham Heath
1823	January	Lynn Districts
		Pinner
	February	Surrey and Sussex
		Kingston District
	November	*Newmarket Heath*
1824	June	*Cambridge and Ely NW District*
1826		*Metropolis Roads*

* South District – Royston to just North of Papworth Everard. Given in the list of 1823 as 'Royston' and also known as Old North Road. 'Royston' in returns to Parliament, 1834 and succeeding years.

Sources: Report of Select Committee on Mr McAdam's Petition, 1823. Minutes of Trusts italicized.

TABLE III : James McAdam's Income, 1823 :

From the Report from the Select Committee on Mr McAdam's Petition, relating to Road Making, evidence of James McAdam, 5th June 1823 :

Mr. *James McAdam*, Examined.

CAN you save the time of the Committee by delivering in any account of the roads which you superintend, and your salary, and other particulars relative to those roads ?—I can ; I have made a list for the purpose of showing the number of miles, the supposed income as well as I have been able to ascertain it ; the present expenditure ; the amount of my own salary when appointed to the road, and the name of the trusts.

[*The Witness delivered in the same.*]

ROADS of which James M'Adam is General Surveyor.

NAME OF TRUSTS	When appointed	Number of Miles.	Supposed Income.	Present Expenditure.	Amount of Salary.	REMARKS.
			£.	£.	£. s. d.	
Hertford Bridge - - -	September 1819	30	2,600	1,800	157 10 –	
Farnham - - - -	November 1819	20	1,500	900	105 – –	
Epsom - - - -	December - 1817	21	3,100	1,800	105 – –	
Cheshunt - - -	October - - 1818	17	4,100	2,200	157 10 –	
Wadesmill - - -	December - 1819	29	3,100	1,800	150 – –	
Staines - - - -	December - 1818	9	1,700	1,210	105 – –	
Huntingdon and Somersham -	December - 1818	20	1,050	750	50 – –	
Oakingham - - -	October - - 1819	20	600	450	52 10 –	
Chichester - - -	April - - - 1820	19	1,300	1,200	105 – –	
Egham West - - -	June - - - 1819	8	1,760	1,040	105 – –	
Godalming - - -	July - - - 1820	20	1,600	1,200	100 – –	
Wycomb - - -	December - 1819	14	2,050	1,250	100 – –	

		Miles		£	
Whetstone	June - - - 1820	8	2,600	1,800	120 - - -
Welwyn	January - - 1820	20	2,000	1,000	100 - - -
Hatfield and Reading	October - - 1820	56	1,800	1,100	105 - - -
Stamford, 5 roads	October - - 1820	42	unknown	3,224	105 - - -
Abingdon and Fyfield	March - - 1822	45¹	900	650	84 - - -
Bury St. Edmunds and Sudbury	June - - - 1821	22	unknown	1,400	100 - - -
Sparrows Herne	March - - 1821	27	2,000	1,200 {first year incomplete}	100 - - -
Stamford Hill	December - 1822	20	12,140		157 10 -
Tonbridge	December - 1822	18	4,000	2,400	100 - - -
Seven Oaks	December - 1822	11	1,900	1,600	60 - - -
Wrotham Heath	December - 1822	17	600	560	40 - - -
Lynn Districts	January - - 1823	94	{unknown, new Act.}	{first year incomplete}	50 - - -
Henley and Oxford	July - - - 1821	27	3,035	2,500	105 - - -
Surrey and Sussex, Kingston District	February - 1823	8	{1 of 9 districts}	1,400	110 - - -
Pinner	January - - 1823	10	570	450	40 - - -
Grantham	February - 1821	15	2,300	1,800	105 - - -
Horkrill²	February - 1822	27	2,800	1,360	105 - - -
Kettering	January - - 1822	12	1,100	800	not fixed.
Boston	February - 1820	28	unknown	2,600	100 - voted 1st year.
Sleaford	February - 1821	26	unknown	1,200	100 - 1st 18 months
Watton	October - - 1821	12	1,200	600	not fixed.
Spilsby	October - - 1822	20	unknown	{Parishesdo the work.}	not fixed.
Higham Ferrers	August - - 1822	12	unknown	640	not fixed.
Miles - -		858		£ - - - £3,479 - -	{Is £4. 1s. 1d. per mile for superintendence, including all expenses.}

¹ This must be wrong. The distance from Fyfield to Abingdon is about 5 miles, and in the Abstracts of Income and Expenditure published from 1836 onward James is shown as Surveyor to Fyfield Trust, of this length (see Appendix One, p. 209).

² Usually spelt Hockerill (see under Essex, Appendix One).

Table IV: Bath Turnpike Trust – Table of Tolls – April 1829

	Full Toll d	¾-Box Gate d
For every Horse or other Beast drawing any Coach Barouche, sociable, berlin, chariot, landau, chaise, phaeton, curricle, gig, caravan, cart upon springs, hearse, litter or other such light carriage except stage coaches, the sum of	4	3
For every Horse or Beast drawing any Stage Coach licensed to carry in the whole inside and outside, not more than Sixteen passengers, the sum of	5	3¾
For every Horse or other Beast drawing any Caravan, tilted Waggon, tilted Cart or other such Carriage (licensed to carry passengers for hire) at the same rate as Stage Coaches carrying the same Number of Passengers	—	—
For every Horse or other Beast (not being an Ass) drawing any Waggon wain or cart or other such Carriage, having the fellies of the Wheels of the Breadth of Six Inches or upwards at the bottoms or soles thereof drawn by not more than four Horses or beasts, the sum of	3½	2⅝
and when drawn by more than four horses or beasts (not being Asses) the Sum of	4½	3⅜
For every Horse or other Beast (not being an Ass) drawing any Waggon wain or Cart or other such Carriage having the fellies of the Wheels of the breadth of four inches and a half and less than Six Inches at the bottoms or soles thereof and drawn by not more than four such Horses or Beasts the Sum of	4½	3⅜
and when drawn by more than four such Horses or beasts the Sum of	5½	4⅛
For every horse or other Beast (not being an Ass) drawing any Waggon wain or Cart or other such Carriage having the fellies of the Wheels of less breadth than four inches and a half at the bottoms or soles thereof drawn by not more than four such Horses or Beasts, the Sum of	5¼	3¹⁵⁄₁₆
and if weighing less than one Ton five hundred weight the Sum of	6	4½
and if weighing less than on Ton five hundred weight the Sum of	4½	3⅜
For every Ass drawing any Waggon, wain or cart or other such Carriage the Sum of	2	1½

228

	Full Toll	¾-Box Gate
For the Horses or other Beasts drawing any Waggon or other such Carriage Laden with one block of Stone (not weighable) the Sum of	s 2	s 1/6
For the same Horses or other Beasts drawing the same waggon or Carriage, every subsequent Time of passing laden with one Block of Stone (not weighable) the Sum of	1/4	1
For the Horses or Beasts drawing any Cart or other such Carriage laden with one block of Stone (not weighable) the Sum of	s 1	d 9
For the same Horses or Beasts drawing the same Cart or Carriage, every subsequent time of passing, laden with one Block of Stone (not weighable), the Sum of	d 8	6
For every Horse or Mule laden or unladen and not drawing the Sum of	1½	1⅛
For every Ass laden or unladen and not drawing the Sum of	1	¾
For every Drove of Oxen, Cows or neat Cattle, per Score and so in proportion for any less Number	6	4½
And for every Drove of Calves, Swine, Hogs, Sheep or Lambs per Score	3	2¼

and so in proportion for any less Number

Source : Bath Trustees' Minutes 24th April 1829.

TABLE VI : Finances of the Exeter Trust, 1823–1836

Year to		Total Income	Balance of Bonded Debt	Total Expenditure
		£	£	£
29 Sep	1823	14,133	54,650	10,872
	1824	14,504	54,600	10,676
	1825	17,663	56,750	12,517
	1826	17,558	66,450	13,797
	1828	21,114	74,400	16,535
	1829	17,271	92,300	12,640
	1830	16,692	95,300	12,270
	1831	17,280	95,800	13,029
	1832	17,440	101,800	12,122
	1833	18,097	102,050	13,537
31 Dec	1834	17,613	109,500	13,369
	1835	19,549	110,200	14,442
	1836	19,085	109,350	13,851

Source : Exeter Trustees' Minutes, Devon Record Office. The figures from 1834 to 1836 do not agree with those printed in Abstracts of Statements of Income and Expenditure of Turnpike Trusts.

Table V: Finances of the Bath Trust, 1830–1853

Year to 30 Sept	Bonded Debt	Income from Tolls	Expenditure on Repairs 'Allowed' or Estimated (E)	Spent
1830	£29,779	£11,061	£7,083	£7,086
1831	43,079	11,282	7,083	7,083
1832	45,629	11,624	7,593	7,156
1833	46,679	10,132	7,443[1]	7,224
Year to 31 Dec				
1834	46,529	10,925	7,443	7,318
1836	46,529	10,692	7,443	7,249
1837	46,179	11,283	7,903[2] GWR works start	7,993
1838	45,529	10,742	7,903[3]	7,993
1839	45,029	13,649	8,403[4]	8,843
1840	43,829	10,762	7,453	8,904
1841	43,829	10,380	7,453 GWR works end	8,543
1842	44,329	10,002	7,000 (E)	7,021
1843	44,329	10,404	6,150 (E)	6,420
1845	43,829	9,636	6,000 (E)	6,038
1846	43,779	9,552	5,800 (E)	5,933
1847	42,829	8,205	5,900 (E)	5,815
1848	42,829	8,727	—	5,702
1849	42,329	8,472	—	5,846
1850	41,829	8,324	5,600 (E)	5,893
1851	41,354	9,036	5,600 (E) Bath City Act	5,653
1852	39,854	8,492	4,520 (E)	4,617
1853	39,854	7,696	4,520 (E)	4,608

[1] £150 'given up' by General Surveyor from sum allowed for repairs.
[2] W McAdam's salary reduced by £200: £250 'extra grant' for repairs.
[3] £450 'extra grant' for repairs.
[4] £950 'extra grant' for repairs.

Sources: Debt and Income from printed accounts of the Bath Trust. Expenditure on Repairs from Surveyor's printed Reports. Both from Somerset Record Office D/T/ba 12.

TABLE VII: Metropolis Roads Commission: Income and Expenditure, 1828–1852

Year to 25th March	Income from Tolls	Expenditure on Maintenance
	To nearest £1	
1828	£75,643	£59,205
1829	79,974	67,513
1830	77,318	60,997
1831	77,604	62,937
1833	76,057	58,315
1834	78,334	62,945
1835	78,894	61,577
1836	79,643	65,920
1837	83,497	68,598
1838	78,629	66,434
1839	73,034	61,355
1840	74,309	63,288
1841	67,475	56,753
1842	68,213	55,961
1843	66,187	58,707
1844	66,310	51,780
1845	66,415	50,389
1846	66,526	55,358
1847	64,914	52,727
1848	63,515	58,727
1849	64,019	59,301
1850	63,539	56,271
1851	65,850	53,732
1852	63,870	56,666

Source: Reports of the Commissioners of Metropolis Turnpike Roads North of the Thames.

Index

ALLEN (coach-master), 152
Allen, John, 129
Alliance Assurance Company, 166
Althorpe, John Charles Spencer, Viscount, 184
Ambler, B, 194
Ambler, Elisha, 15, 187, 194
Amiens, Peace of (1801), 28
Annual Continuation Acts, 10
Armstrong, W A, 86
Arrington, 13, 89
Automobile Club of Great Britain (*later* Royal Automobile Club), 204
Aylesbury, 12
Ayr, 35
Ayrshire Bank, 23

BARING, Sir Thomas, 106, 109, 138
Barnet Hill, 180
Barnstaple Trust, 139
Bath Turnpike Trust, 9–11, 13, 119, 121; toll charges, 15, 228; and road construction, 33, 116, 121–4; JLM and, 60, 77, 80, 98, 116, 121–6, 129–30, 145, 153; and Wingrove, 110, 121–9; JLM's report to, 123–4; JLM appointed surveyor, 131; William becomes deputy,

131, 139, 145, 152–5; 1840 accounts, 146–9; and new Melksham road, 150–51; 1829 Act, 152, 154; controversy with Black Dog over Warminster road, 153; and Great Western Railway, 154–5; 1851 Improvement Act, 155; finances (1830–53), 229
Berkhamsted, 13
Black Dog Trust, 80, 131, 145, 153, 155
Black Horse hill, 53
Boord (coach-master), 152
Botham, George, 17
Bradford-on-Avon Trust, 150
Brassey, Thomas, 48
Bridgewater, Francis Henry Egerton, 8th Earl of, 12, 14
Bridgwater (Somerset), 32, 151
Bristol, 20, 42–3, 56, 154–5
Bristol Turnpike Trust, 8, 42, 43, 48; costs and debts, 49, 53; JLM's surveyorship of, 49–52, 55–7, 60, 76, 98, 105, 131; JLM's improvements, 52–3; 1819/20 Act, 59; JLM's resignation and reappointment, 62–4; Loudon succeeds to surveyorship, 64, 155; James junior with, 76;

Index

Index

Holyhead Road Commissioners, 20, 172–80

Hood, Thomas, quoted, 67–70

Hooley, E Purnell, 206

Hooper (Bristol surveyor), 154

Horne, William, 17

House of Commons, debates public money for JLM, 94, 109, 111–12; and Wingrove, 128–9

Select Committee: 1811, 31, 37–9; 1819, 16–17, 54, 58, 104–5, 113–14, 126, 162; 1820, 107, 162; 1823, 27–30, 52–3, 71–2, 95, 100, 102, 107, 109–11, 127, 226; 1825, 166; 1831, 192

House of Lords, 1833 Select Committee, 185

Hume, Joseph, 111, 164, 181

Iddesleigh, Stafford Henry Northcote, 1st Earl of, 159

Inglis, Harry R G, 204

James (of Radstock; Bath trustee), 121

Johnson, Charles, 104, 106, 108

Johnson, Samuel, 15

Kendal (Westmoreland), 33, 35

Kidderminster Trust, 79

Kilby, Sarah see McAdam, Sarah

Kinglake, W, 127

Kingsbridge Trust, 139

Kingsdown Hill (near Bath), 17, 150, 152

Knatchbull, Sir E, 111

Kneesworth and Caxton Trust, 86, 87, 91, 97, 191

Lamb, George, 185

Lancashire, 75, 77

Launceston Trust, 97, 145

Lester, William, 110

Levy, Lewis, 15, 166, 187, 194

Lewes (Sussex), 57

Liverpool, 42

Liverpool, Robert Banks Jenkinson, 2nd Earl, 75

Lloyd George, David, 205

Local Government Act, 1895, 201

Local Government Board, 201

London, road services, 15, 159, 161; James' surveyorships in, 77–8, 100–1, 159, 167–71; tolls in, 159, 161, 165, 169, 188–90; consolidation of turnpikes in, 159, 162–5, 168; road improvements in, 167, 169–71, 194; macadamizing in, 169; railways in, 187; increase in size and traffic, 188–90; see also Metropolis Turnpike Roads

London and Birmingham Railway, 187

London and Bristol Rail-Road Company, 61–2

Lonsdale, William Lowther, 2nd Earl of, 162–4, 186

Louis Philippe, King of the French, 186

Lowther, William see Lonsdale, William Lowther, 2nd Earl of

Lucas (Bristol road contractor), 63

Lyon, Mr, 184–5

McAdam family, surveyorships, 75–7, 85, 209–14

236

Index

McAdam, Ann (*née* Dey; uncle William's wife), 24, 41

McAdam, Anne Charlotte (*née* De Lancey; JLM's 2nd wife), 112, 130

McAdam, Christopher (JLM's grandson), as surveyor of Exeter Trust, 64, 80, 119, 132, 136, 139; career and other surveyorships, 71, 98, 139, 142, 145, 155, 209–14, 224; favoured by Truro Trust, 97; earnings, 100; and Bath Trust, 119; Exeter contract terminated, 144–5, 201

McAdam, Christopher Kilby, 72

McAdam, Georgina Keith (JLM's daughter), 41

McAdam, Gilbert (JLM's uncle), 215

McAdam, Gloriana Margaretta (JLM's daughter), 41

McAdam, Gloriana Margaretta (*née* Nicoll; JLM's first wife), 26, 130, 215

McAdam, James (JLM's father), 23–4

McAdam, Sir James Nicoll (JLM's son), and decline of roads, 2, 191–2; and road construction, 3; and Metropolis roads, 7, 77–9, 100–1, 159, 164, 166–71; and administration of road system, 8, 11, 12, 15, 96, 158, 159, 179, 183–5, 193, 200; early business interests, 41, 71, 73; advocates county surveyors, 47; and road

labour and wages, 54–5; as Epsom surveyor, 60, 73, 182–3; other surveyorships, 72–4, 77, 85, 155–6, 209–14, 223–7; and Cheshunt Trust, 86–7; and Wadesmill Trust, 87–91; and Kneesworth and Caxton Trust, 91; and St Albans Trust, 93, 172–6, 179; and accountability, 96–7; on demands of work, 98; salaried assistants, 98; earnings and expenses, 99–101, 164, 191, 226–7; and 1819 Select Committee, 104; achievements, 158, 159, 185, 194; presents turnpike returns to Parliament, 158; attacked over Holyhead road, 172–6, 179; and Duke of Richmond, 181–2, 184, 186; political activities, 183–5, 200; visits France and Channel Islands, 185–6; knighthood, 186; and railways, 186–7; and London traffic, 188–90; and steam vehicles, 192; and proposed monument to father, 193–4; death, 194

McAdam, James (Sir James' son), 72, 191–2, 194, 209–14

McAdam, Jane (*née* Pickard; William senior's wife), 71, 215

McAdam, John Loudon, and road construction, 1–3, 5, 32-9, 52, 114-15, 126-7, 161; and piecework labour, 14; and turnpike trusts, 20, 44–5; birth and

237

family, 23–4; in America,
24–6; first marriage, 26,
215; Deputy-Lieutenant of
Ayrshire, 27; early business
difficulties, 27–8; as Inspector of Roads, 29; on
surveyors, 36, 44, 46–8;
as General-Surveyor of
Bristol Trust, 41, 43–4,
48–52, 60, 62–3, 76, 131;
improving means, 41–2; on
trustees, 44–6; and labour
for roads, 54–5; Bristol
contract renewed, 55–7; and
Chichester, 57; consulted by
other trusts, 58–60; family
associations, 60–1, 71–3,
112; and London & Bristol
Rail-Road Co., 61–2; resignation and reappointment
to Bristol Trust, 63–4;
other surveyorships, 64, 74,
80, 209–14, 223–4; Ches-
hunt Trust recommends, 87;
opposition to, 93–5; public
money for, 94, 102, 106–12,
117; accounting and management system, 95, 97–8, 115–
17, 140; advocates centralized road system, 96, 105,
205; earnings and expenses,
99, 102, 112, 131; and 1819
Select Committee, 104–5,
114; and 1823 Select Com-
mittee, 109–11; reputation,
112; name adopted for
process, 112; and Bath Trust,
119, 122–6, 129–30, 145;
appointed surveyor of Bath,
131; second marriage, 130;

and Exeter Trust, 136–8;
and Bath-Melksham road,
150–2; visits to Scotland,
154; death, 154–5, 193;
on London roads, 161; in St
Albans road conflict, 176;
offered knighthood, 186;
proposed monument to, 193–
4

Works: *The Management of
Trusts*, 44; *Narrative of
Affairs of the Bristol District
of Roads*, 63; *Remarks on the
Present System of Road
Making*, 31, 44, 47

McAdam, Kilby, 215

McAdam, (John) Loudon (JLM's
son), and road construction,
3; birth, 28, 41; and Bristol
roads, 43; as Joint General-
Surveyor of Bristol Trust,
51; succeeds father as
General-Surveyor, 64, 76;
assists father, 71, 73; other
surveyorships, 73, 76, 81,
155, 209–14, 223–4; travels,
99; earnings, 102; and
Shepton Mallet Trust, 131–
2; in West, 194; and
Sheffield Trust, 201

McAdam, Sarah (*née* Kilby;
Gilbert's wife), 215

McAdam, Susannah (*née*
Cochrane; JLM's mother),
23

McAdam, Wilhelmina Hannah
see Shaw, Wilhelmina
Hannah

McAdam, William senior (JLM's
son), and road construction,

3; and turnpike trusts, 9; returns from America, 27; manages Muirkirk Iron Co., 28; manages British Tar Co., 41, 60, 71; and Bristol roads, 43; assists father, 60; with Exeter Trust, 64, 119, 132–40, 142; marriage, 71, 215; death, 72, 143, 155; roads career and surveyorships, 72–3, 76–7, 80–1, 93–5, 209–14, 223–4; and Truro Trust, 82–5, 201; earnings and expenses, 84–5, 98–101, 139; and opposition to family methods, 93–4; Estcourt criticizes, 94, 111; management and control, 95–7; and JLM's claim for public money, 110; and Bath Trust, 122, 125, 153; with Black Dog Trust, 131, 153; in Devon, 139

McAdam, William junior (William senior's son), 43; career, 71, 81, 131, 145, 155, 194; earnings, 101, 154; at Bath, 119, 131, 139, 145, 152–5; at Exeter, 137; on railways, 154–5; death, 155, 199; surveyorships, 209–14, 224

McAdam, William (JLM's uncle), 24–5

McAdam Watson & Co., New York, 26–7

Macaulay, Thomas Babington, 1st Baron, 159

McConnell, John, 74, 99

Macintosh, Charles, 1

Manvers, Charles Herbert Pierrepont, 2nd Earl, 194

Melksham Trust, 94, 150, 152

Melville, Robert Dundas, 2nd Viscount, 60

Metropolis Roads Act, 1826, 159, 163, 167

Metropolis Turnpike Roads, Commissioners for, James as General Surveyor for, 74, 77–9, 100–1, 159, 164, 166–71, 194; formed, 159, 164, 166–7; character and financing of, 164–6, 171; work and improvements by, 168–71, 194; and Turnpike Returns Act, 184; and railways, 186–7; and increased London traffic, 188–90; attacked, 190–1; income and expenditure (1828–52), 231

Miller, George Layng, 130

Mitchell, Joseph, 74, 116–17

motor cars, 204–6

Mott family (of Much Hadham), 13

Much Hadham, 13

Muirkirk Iron Company, 27–8, 42

New York, 24–7

New York Chamber of Commerce, 24

Newmarket Heath Trust, 12, 182, 191*n*

Nicoll, Glorianna Margaretta *see* McAdam, Glorianna Margaretta

Norris, Henry, 92–3